The Othe

Sophie Flynn is a Cotswolds based psychological thriller author with an MA in Creative Writing from Oxford Brookes. Alongside writing, Sophie is the Managing Director at Jericho Writers.

Also by Sophie Flynn

Keep Them Close
If They Knew
What Stays Unsaid
The Other Miss Jones

SOPHIE FLYNN

THE OTHER MISS JONES

hera

Penguin
Random
House

First published in the United Kingdom in 2025 by

Hera Books, an imprint of
Canelo Digital Publishing Limited,
20 Vauxhall Bridge Road,
London SW1V 2SA
United Kingdom

A Penguin Random House Company

The authorised representative in the EEA is Dorling Kindersley Verlag GmbH. Arnulfstr. 124, 80636 Munich, Germany

Copyright © Sophie Flynn 2025

The moral right of Sophie Flynn to be identified as the creator of this work has been asserted in accordance with the Copyright, Designs and Patents Act, 1988.

All rights reserved. No part of this publication may be reproduced or transmitted in any form or by any means, electronic or mechanical, including photocopy, recording, or any information storage and retrieval system, without permission in writing from the publisher.

No part of this book may be used or reproduced in any manner for the purpose of training artificial intelligence technologies or systems. In accordance with Article 4(3) of the DSM Directive 2019/790, Canelo expressly reserves this work from the text and data mining exception.

A CIP catalogue record for this book is available from the British Library.

Print ISBN 978 1 80436 674 5
Ebook ISBN 978 1 80436 672 1

This book is a work of fiction. Names, characters, businesses, organizations, places and events are either the product of the author's imagination or are used fictitiously. Any resemblance to actual persons, living or dead, events or locales is entirely coincidental.

Printed and bound in Great Britain by Clays Ltd, Elcograf S.p.A.

Look for more great books at
www.herabooks.com
www.dk.com

For Rosie,

My first reader, character inspiration, and unpaid researcher.

PROLOGUE

She had been dreaming of this moment for years, but now it was here, she was terrified. His body was so close that if she reached out, she could touch him. Feel his cold, dead skin against hers. She shivered. This wasn't how things were supposed to go.

He'd been screaming at her. Grabbing her by the hair until chunks tore from her skull in bloody patches. He'd let go only to wrap his thick, calloused hands around her neck instead.

Crunching; crushing.

But just as quickly as the attack started, it stopped and then he was on the floor in front of her. Blood pooling from the back of his head.

Things had got so out of hand so quickly. Wasn't this what she wanted when she came here? Him gone, her free. Except she wasn't free. Now she was trapped forever by what happened here.

Next to her, the baby screamed. She looked down – no. Not the baby.

Someone else was screaming. No, not screaming. Shouting. The familiar, violent sound of someone who had not got what they wanted. A warning. The shout would soon turn to violence; she knew the sound so well. It had been the one consistent thing in her life.

This wasn't supposed to happen. He was supposed to keep her safe.

It was too late now for her plan to work. She knew that she could never run off into a brand-new life; her baby with a daddy, her with a protector. Everyone who ever promised to protect her had left in the end.

She looked around her – the darkness rearranged itself as the whooshing in her ears unbalanced her. More men would come. She was sure of this and when they did, they would do more than crush her neck between their dirty hands. She stopped – thinking, trying to remap her plans and find an escape. But there was nothing. She had used up all her chances, all her favours. There was nowhere left to go – no one left to turn to.

She didn't trust the police, never had. But it wasn't about her now. She looked down at her little girl, her perfect, innocent baby girl. She wasn't crying any more. Maybe she never had been. She gurgled and smiled up at her mother with her big brown eyes. Perfectly content. *I have to keep you safe.*

She took her phone from her pocket.

'Police,' she whispered through her crushed throat.

'You're through to Warwickshire Police. What's your name and your location?'

Her throat clamped shut. She heard something. Footsteps. Running away or towards her, she couldn't tell. A threat. It was coming closer.

'Hello? I need your name and location, please.'

The voice on the other end of the line was warm and it made something rip inside her heart. Were people still warm? Did people like this woman really exist? She imagined the woman talking to her from somewhere soft, somewhere calm. Somewhere safe. The baby's warmth radiated through her, a soft bundle that she had to protect.

'I'm scared...' she whispered. 'Please help me.'

'I'm going to help you but first I need your name. How about I tell you mine, then you can tell me yours?' the woman said, a simple, kind offering that neither of them knew then would change the course of both their lives forever. 'My name is Sally Jones.'

PART ONE

CHAPTER ONE

FIVE YEARS LATER

It's the look on the Inspector's face that does it. Her kind but slightly embarrassed smile as she tells me, 'Honestly, Sal, you know as well as I do that it's for the best. You've not been yourself lately and we can't have you getting so… emotional. Not here. Not in your role. I know you want to help everyone, but you also have to look after yourself.' She pauses for a second as my cheeks flame, then softens her tone. 'You've not given yourself any time to grieve your dad's death and I think it's impacting you more than you know.' That's when the tears come, hot and fast out of nowhere.

'It's only for a few weeks,' Inspector Holden says in that calm but a little bit reserved tone she takes when any of us show too much emotion. 'Hopefully,' she adds with a little bit more uncertainty.

'Hopefully?' I say through a pathetic sniffle.

'Well, you know how these things can go. HR have said not to put any stringent time frames on it.' She rolls her eyes, nodding towards the other end of the corridor where the offices of the notorious Human Resources department sit. I slump down in my chair. I can't believe my name has become a file in their team's workload. Of all the people you'd expect to be called into the Inspector's office and forced to take some time off for the good of their 'mental health', it would absolutely not be me.

'Okay,' I say, wiping the snot from my nose in an attempt to pull myself together. I need to be the secure, in-command person Inspector Holden is used to. From inside her glass fronted office I can see Jane, one of the trainees, waiting to come in. I bet she'll replace me. They'll tell her I can 'no longer cope' with the stress of the job – but they know she'll manage just fine. Bloody hell. It's enough to force my face into a neutral expression as I wipe the last remaining tear from my cheek.

'Well,' I say, all business again now. 'I suppose that's that, then. I'll wait to hear from you.'

The Inspector nods, relief washing over her face that I'm not going to continue this outward display of emotion. With that, I turn and strut out of the office, giving Jane a sharp glare as I pass.

—

A few hours later, back at home, I take cold, slimy chicken out of a packet and slice it into two-centimetre-thick pieces before throwing it into the hot pan. It sizzles and spits fat onto my naked arms.

'Pissing hell,' I say, running the cold water tap and dunking my arm in. Who would have thought a TikTok recipe would be so bloody hazardous?

I flip my phone screen back to the video and the high-pitched, overly happy voice of the 'content creator' blares out again.

'Cook my "Marry-me chicken" with me for your man tonight…'

I mute it because the phrase 'Marry-me chicken' makes me feel like some sort of wannabe tradwife and I'm not actually going to be Nathan's wife for two more weeks – so it feels somewhat like tempting fate. But still, I need something to take my mind off the fact I've been banned from work and my whole team are going to know I'm not considered 'stable' enough to deal with the general public's emergencies right now.

'Ten sun-dried tomatoes...' I say to myself as I read the recipe from the screen, 'chopped into "bite-sized pieces"... I can do that.' I take out the jar I got earlier from Waitrose and fish around in the oil before getting a hold of the slimy things and slicing them up on the chopping board. This is fine. Totally distracting. Maybe I should do it more often. I could become the type of woman that cooks from scratch! Who needs a career anyway?

Once the pasta is on to boil, I stand back and survey the wreckage of the kitchen. There are globs of cream on the side, oil pooling from the tomatoes on the chopping board, and the gross white bits I chopped off the chicken discarded by the sink.

I can hear Dad's voice in my head, '*Wash as you go, Sally! Wash as you go.*' It was always Dad that cooked; I guess he had no choice given that Mum left before I was even in school. She was never calm enough to stand for hours sorting my dinner; even when she was around, it was always Dad's job. You'd think the time I'd spent watching him cook for me whenever he wasn't on shift would have rubbed off and I'd be able to put together a decent meal without much fuss, but unfortunately his skills didn't transfer. He's been dead for six whole months and doing anything that reminds me of him has been impossible. But now, given work doesn't want me there, I guess it's as good a time as any to learn to cook while I 'process my grief'. Whatever the hell that means. Looking at the mess I've made in the kitchen, it doesn't seem worth the effort. It's going to take me forever to tidy this up.

But then again, I've got all the time in the world now.

The front door goes as I'm finishing the washing up. I grab the pinny I bought earlier from the hook on the back door and tie it around my waist, then scrape my hair back into a ponytail to try and appear a little bit put together.

'In here, honey!' I call out in a sing-song voice.

Nathan walks through the door and then takes a step back at the sight of me, and the kitchen.

'What the...' His face screws up as he surveys the boiling pot behind me, the kitchen table set for two (with a candle in the middle and everything) and lastly, my brand-new pinny with the phrase 'Queen of the kitchen' scrawled in a loopy handwritten font on the front. B&M's finest.

'Do you like it?' I ask, tucking my hair behind my ear.

He says nothing for a moment, then steps forward and pulls me into a bear hug. 'You're bloody nuts,' he says, then lets out a warm laugh that vibrates through me.

—

We sit opposite each other at the table, the candle flickering between us. I've served up the pasta into two great big white bowls and, I must admit, it does taste pretty amazing. Maybe I will become a proper cook after all.

'If this is what happens when you "take some time for yourself" then I'm in full support,' Nathan says, taking a huge forkful of pasta and shoving it in his mouth. He chews, then swallows.

'You know me,' I say. 'Anything to "avoid my grief",' parroting the phrase I've already told him was used against me at work today, while rolling my eyes.

He cocks his head to the side, no longer laughing. 'Holden might have a point, you know, babe,' he says. 'You do need to look after yourself, not just everyone else.'

I nod, trying not to snap back at this and instead look down at my bowl, the thick cream sauce suddenly feeling uncomfortably heavy in my stomach at the idea he agrees with my inspector.

'Plus, you can't keep cooking like this for me. I won't fit into my wedding suit,' he says, kicking my foot gently under the table, trying to make me laugh. 'Where'd you get the recipe for this anyway?'

I look up and purse my lips.

He laughs. 'God, not TikTok?'

'Take the piss all you want, I'm down with the kids.'

'Please never say that again,' he says, groaning.

'Well, maybe not the kids. But I'm down with the tradwives. Guess what the name of the recipe was?'

He tilts his head to the side, stroking his thick, dark brown beard. 'Hubby's favourite chicken?' He says in the American accent he affects when he overhears me watching the bonkers TikTok videos I've become slightly obsessed with over the last few months while sleep has evaded me. I'm so far from a housewife it's ridiculous, yet there's something so comforting about watching these American housewives detailing their day-to-day lives like something from the Fifties.

'Even better,' I say. 'It's called "Marry-me chicken".'

He bursts out laughing and stands up, pulling me from my chair then guiding me to his side of the table where he sits me back down on his lap. I lace my fingers around the back of his neck as he pulls me closer.

'Well, I'd say the recipe has worked its magic then, hasn't it?' he says. 'As that's exactly what I'll be doing in two short weeks.'

Before I can roll my eyes at his cheesy line, he pulls me in for a kiss.

Once we part, I go back to my side of the table and polish off the last few tubes of penne.

'You know, you don't need to worry about me, right? I *am* going back to work.'

He nods. 'Yes, Sal. While I do love the idea of you spending your days relaxing before cooking me up a delicious meal, I'm very aware you aren't going to transform from someone who thinks there's nothing more "fun" than a murder in the middle of a nightshift, to a "tradwife" overnight—'

'Not just overnight,' I say, placing my fork down. 'Not ever.'

'I know!' he says, eyes wide. 'Do you want to talk about what happened today?'

I sigh then push my plate away. 'What's there to say? They don't want me on shift any more.' I had called Nath on his way home from work and told him the embarrassing news and said we'd talk about it later. But now later is here and I still don't want to.

'It's just for a couple of weeks—'

'It might be more than two weeks—'

He waves his hand. 'That's what I said—'

'A *couple* of weeks is two. Holden said she hoped it would be a *few* weeks. That's at least three.'

He doesn't respond for a minute, just looks at me with a smile tugging at his lips.

'Maybe Holden's right,' he says, not meeting my eyes. 'You have been getting pretty emotionally invested in your calls lately, and you haven't really given yourself much time to grieve your dad—'

I scoff. 'What difference will grieving make? It won't bring him back.'

Nathan smiles, sadly. He loved Dad too, who was as much of a father figure to him as Amanda, his mum, was a mother to me.

'I know you think I'm ridiculous,' I say.

He shakes his head. 'I don't. I get it. You don't want to be on enforced leave. I wouldn't either. But it's just a couple of weeks; it's going to go by so fast and then it'll be the wedding and then your "enforced holiday" will be over and you'll be back at work like normal.'

I nod, trying to believe what he says, to be positive like he is that I'll simply be able to see these few weeks of doing nothing as a 'fun holiday' and not let my grief completely spiral into a place I won't come back from, and that things will go back to normal as if nothing ever happened. The thing is, I know my inspector is right. I've not dealt with the grief of losing Dad; I've hidden away from it in the darkness of my work bunker, listening to other people's lives fall apart and trying to fix them instead. I'm not able to admit this to Nathan and he wouldn't hear it even if I could. He always sees the best in me; it's one of the reasons I love him so much, but it does mean there's an awfully long way to fall if he realises the truth that perhaps I'm not the great fixer I once was and now I'm the broken one instead.

'It wouldn't be so crap if you could take some time off, too,' I say now to Nathan as he collects up our dishes and stacks them in the dishwasher behind me. 'Can't you even take a few days?'

'No chance,' he says, shaking his head. 'I've already told you this, babe. You know we're in the middle of this big op, I'll be working overtime non-stop over the next few weeks. Everyone's had their leave cancelled; we're lucky they're even letting me take the two days off for the wedding.'

I murmur agreement and let out a big sigh. I know he's right. There's no way in hell he's going to be able to chill out with me at home while this op is going on.

'I think I'll go mad,' I grumble, as he kisses me on top of the head.

'Most women would be quite pleased to have a couple of weeks of "me time" ahead of their wedding.'

'Yeah, well,' I say, sounding like a petulant teenager. 'I'm not most women.'

'No, Sally Jones, you absolutely are not.'

I watch his back as he washes up, the muscles in his shoulders hard against his black t-shirt. Nathan has always kept fit but in the last year he's upped his game. I poke at my own body, the lack of muscles severely apparent though I'm less pudgy than I was six months ago. The grief diet, I believe they call it.

Though Nath and I are both in the police – like our parents were – our jobs couldn't be more different. He's always on the move as Detective Chief Inspector of the Serious Crimes Unit, whereas my role rarely involves me moving from my seat, as a call handler in the Command Control Centre – known as the OCC, or as Abigail, my 'kind-of' niece calls it, 'the place where they answer the 999 calls'. She says this proudly and I have to stop myself from telling her I'm more than someone who answers the phones like a secretary; that I make life-changing decisions every single day about who gets the help when they need it most and who's left to hope for the best. After all, she is only five. The biggest difference is that I effectively work in a bunker – it has to be secure if there's ever a bomb dropped on us, for example – with no windows to the outside world and see the crimes through multiple TV screens and the details written in black and white on my computer, whereas Nathan has been face-to-face with the criminals and victims every day throughout his career.

I prefer a bit of detachment. It's much easier to make a decision about which victim gets help from the police immediately and which need to wait when you can't see their faces. But perhaps that's where I've gone wrong – imagining the lives of the people I can't see and rewriting them into fairy tales where I'm the hero. The job has changed me over the years; perhaps that's why I'm finding it so hard to cut myself open and let the grief pour out for my dad. Heroes aren't supposed to fall apart.

'Wine?' Nathan asks, already pulling the bottle from the fridge and pouring me a glass. It's rare we have an evening together like this. Usually one of us is on shift and sometimes we can go weeks without sitting down together for a proper meal. I suppose that'll be a positive change while I'm off work, at least. Though by the sounds of it, Nathan will barely be home if this investigation he's on doesn't slow down soon.

I say yes to the wine, then keep scrolling, my eyes not taking in the endless stream of photos of foamy lattes, sunsets and cocktails that Instagram serves up to me. An email pings up and I swipe to read it.

SUBJECT: IMPORTANT: PROPERTY 161 BELLEVUE TERRACE

Dear SALLY JONES,

> This is a formal communication from LAYTON & HARDING LETTINGS.
>
> You have been contacted by post three times on the dates 1 MAY, 2025, 10 MAY, 2025, 13 MAY, 2025 and failed to respond. As per your lettings agreement, as you are now TWO MONTHS in arrears, we have the right to evict you from the property address 161 BELLEVUE TERRACE, WESTON-SUPER-MARE. Please make payment by FRIDAY 31 MAY, 2025, or we will remove you from the property. If you have made payment within the last three days, you may disregard this letter.

'Shit,' I say under my breath, reading the email again.

'What's up?' Nathan asks, looking over my shoulder.

I flick off the email and put my phone face down on the table. 'Nothing,' I say, giving him a tight smile. 'Just some horrible news story.'

He raises his eyebrow, not knowing me to ever be particularly fazed by the news – after all, our jobs bring us into contact with the worst news every day – but he doesn't pursue it and nods towards the living room.

'Find something on Netflix,' I call through. 'I'll be in in a minute.'

I get the email back up, reading it again with a knot in my stomach.

Without thinking it through that much I forward it.

Are you okay? I write. *Can I help with anything? I'm worried, let me know. S x*

I listen to the whoosh of the email flying over and wait for a reply. Maybe someone does still need my help after all.

CHAPTER TWO

Nathan started his shift at four a.m. this morning and I know this because, despite it being a Sunday with no shift in sight for me, I was wide awake. I've worked shifts for fourteen years and I don't think my body will ever recover from being forced out of bed at irregular hours. So many articles I've read over the years warn about the impact of shift work on women in particular, and I've rolled my eyes and ignored them. Like yes, sure, it's probably not ideal to live like this but how else do you suggest we provide twenty-four-seven emergency cover? No one wants to do our jobs but they all bloody well should otherwise the whole country would be in total chaos. As I rant about this in my head, I can hear my dad's tut and roll of his eyes. He thought it was so funny when I got on my high horse when I first joined the police, an idealistic twenty-year-old with ideas of grandeur, following in his footsteps but certain I'd do it differently to him.

Though it's been six months since he died, my life is narrated by his gruff voice in my head most days, like he's still here. He'd be so disappointed that I've been forced into taking leave; he'd never show it, mind. He'd tell me what everyone else does – not to worry and it'll all be right in the end. But deep down, I know he'd be as horrified as I am that his Sally, his golden girl, can't cope any more.

I make another cup of coffee, my third of the morning, and check the time on my phone. Just gone ten thirty. My phone bleeps and I hope it's a reply to my email last night but it's just Amanda, Nathan's mum.

> Lunch at 2 today x

I quickly reply, reminding her Nath won't be there but I'll see her soon. Whenever our shifts allow it, Sunday lunch at his mum's has always been a staple of our week. She retired five years ago and for the first time in her – and Nath's – adult life, was able to be the sort of mum you read about in magazines. She went from being a WPC – back in the day when it was still called that – as a bolshie eighteen-year-old, to working her way up to Assistant Chief Constable of Warwickshire Police, and then having to fill her days with gardening and cooking us roasts. But I think she's pretty happy now; a bit more settled, at least, and I'm glad to have our lunch in my diary today to give me something else to focus on instead of worrying about work and the lack of reply to that troubling eviction email.

–

I walk up the path to Amanda's house, a homely detached red-brick property in a village ten minutes from our place in Stratford-upon-Avon, and check my phone again. Still no reply. I put it away as Amanda opens the door.

'Sally!' she says, her face brightening into a smile.

I have to say, being retired certainly agrees with her. She's always been attractive, but when she was working her forehead was mostly creased into a series of sharp lines that made her look so terrifyingly severe it earned her the nickname 'Ice Queen' on the force. I'd never tell her that, of course. Though I'm one hundred percent sure she is well aware of the nickname and quite enjoyed it. She always said to me you can't spend time worrying what's said about you as a woman in the police; everyone will assume you're half as good at your job as any of the men even when you're a million times better.

'Hey,' I say back to her as I walk into the house. 'Here you go.' I hand her the dessert I've brought with me – sticky toffee pudding – that I picked up from Marks on the way over and she eyes it suspiciously.

'Why did you bring this?' she asks.

I shrug. 'Thought it was the sort of thing you were supposed to do. You cook the food; the guest brings the pudding.'

'Guest?' she says, her eyebrows crinkled above her ice-blue eyes.

I laugh. 'Well, you know what I mean.'

'You'll be baking soon if they don't get you back to work,' she says with a frown before she turns, and I follow her through into the kitchen.

'Sally!' Abigail jumps up from the table and runs towards me, her arms outstretched.

'Hello, trouble,' I say, picking up the not-so-little-any-more girl. 'Cor, you're heavy!' I say overdramatising my struggle to lift her as she giggles in my arms. I wave hello to Chloe and Matt, Abigail's parents and our long-time family friends, who – as usual – look absolutely knackered.

'No Nathan today?' Matt says and I shake my head. He looks disappointed.

'He's on shift,' Amanda says, and I detect a hint of wistful nostalgia for her own hectic career when a Sunday afternoon off was nothing but a dream.

'Is he catching the bad guys?' Abigail says, now safely sat on my lap at the table, as she twists my necklace around her thin little fingers.

'Yes, darling,' Amanda replies. 'He's out there so we can stay in here safe and sound.' Abigail nods, her big brown eyes round and innocent, like the bad guys are something from story books that could never touch her cosy life, and I hug her tighter to me, remembering how close this little girl came to living such a different life, where all she knew was bad guys.

'Why don't you go and sit in the garden until I call you in,' Amanda says. 'It won't be long but it'd be much easier to get everything ready if you weren't all under my feet.'

Matt chuckles, totally unoffended by Amanda's brusque manner.

'Yes, ma'am,' he says with a mock salute as she rolls her eyes and flips the tea towel at him. Matt takes Abigail's hand to lead her out, but she stomps her small feet and shouts for her mum.

'I'm coming, sweetheart,' Chloe says, shrugging at me with a little smile.

They adopted Abigail as a baby but she still has attachment issues, as Chloe calls them, and if her mum isn't in sight for even a second, Abigail quickly goes from being the sweetest kid around to a little devil. Amanda is the only other person who seems able to coax her out of these tantrums – although they're not blood-related, she's always acted as Abigail's grandma. It's a funny patchwork family, but it's one that works.

In the garden, once Matt has brought me a cold glass of wine and Abigail is safely amused on the swing set, I sit back in the chair and enjoy the feeling of warm sun on my face.

'When are you back on shift?' Chloe asks me, instantly breaking my sense of calm.

'Oh, not for a few weeks—'

'Weeks?' Matt says with a raised eyebrow. 'Don't tell me you're actually taking some holiday!'

'Um yes, kind of...' I say, watching the drips of condensation travel down my wine glass. 'I've been put on leave for a bit...' I stop, not knowing how to tell them what's happened. I look up at Chloe's kind, concerned frown. I'm making it sound far worse than it is, so I wave my hand and finish up. 'They've asked me to take some leave. I've been a bit stressed at work, since Dad died. It's no big deal, happens a lot in our jobs. They don't want anything to impact performance...'

Matt whistles through his teeth, like this is actually an incredibly big deal.

'Oh, darling,' Chloe says before taking a sip of her wine. 'I'm so sorry, Sal. You must feel awful.'

I shrug, just as Matt asks. 'Did something happen then? Did you do something wrong?'

And for a moment I think he's asking me this directly, as if he needs to understand what went wrong in my life to bring me to this point, before I realise he means at work.

'Well...' I start, with no idea how to carry on. 'The thing is, you have to make some difficult decisions in my job sometimes and you don't often have a lot of time to think about them...'

Luckily, I'm interrupted by Amanda who has come out onto the patio with a bottle of wine in hand. She smiles broadly at Chloe and Matt and takes over.

'It's all nonsense,' she tells them. 'Our Sal hasn't done a thing wrong. But it's all bureaucracy nowadays. With the joke of our PCC lording it up in his Tory-Red trousers from the safety of his mega-office as if he has the slightest idea what real policing is like.'

'So, is it this person, this PCC that's said you need to take some time off, Sal?' Chloe asks.

I scrunch up my nose. 'Oh god, no,' I laugh. 'He wouldn't concern himself with mere mortals like me. They think I've not been working "to my usual standards" lately, got a bit caught up in the outcome of my calls. Nothing serious. I've not "dealt with my grief" and have let it affect my work, apparently,' I say, sipping my wine before remembering I'm driving later and pushing the glass away. 'But you know, it's been stressful.'

'Sounds awful,' Chloe says sympathetically. 'You'd think they'd understand you having a few bad weeks given you've just lost your dad.'

'Sounds like a piss-take,' Matt adds. 'There's no one with more integrity than you, Sal. There's no way you've done anything wrong. Whether you were upset over your dad or not, there's no one I'd trust more with making a tough decision than you.'

I smile, trying to let Matt's words sink in. But of course, Matt thinks I'm an angel. After all, I'm the reason they have Abigail; I'm the reason she was saved.

'Too right,' Amanda says. 'And I've told Sal already she has absolutely nothing to worry about. This "enforced leave" thing is just HR covering their backs. You just need to sit tight,' she says to me. 'It'll all be over before you know it.'

'Can we do anything to help?' Chloe asks as Abigail comes running over and jumps up onto her lap.

'Yeah, we could write a letter in support of you. You know, tell them what sort of person you really are,' Matt adds.

'Are you in trouble?' Abigail asks, looking up at me again with those big, worried eyes.

'No, darling,' I say while the voice in my head says I might very well be.

'Sally is just taking a little holiday so she can go back nice and rested,' Amanda says. 'Like you do after a long term at school.'

Abigail nods but looks unhappy about it, as if she knows Amanda isn't quite telling the truth. 'If you're in trouble that's not fair,' she says, unsure. 'Because you're a hero.'

Chloe and Matt smile at their daughter with doting eyes and Amanda laughs softly from behind me.

To Abigail, I will always be a hero. She was just a baby when I first heard her cry as her birth mother begged me for help. It was the call that earned me my Chief's Commendation Award just a few months into my policing career; it's the reason I became the golden girl of the department. It's also the reason I've got myself into such a mess now. That night coloured my view of victims and perpetrators forever. I fully believed the caller, Natalie Pierce, that she was the one in danger. I helped her to hide Abigail and disappear – telling her what a good mum she was. I thought she was in danger, that she was the victim but it turned out she was a violent psychopath. I later learned how she killed Abigail's birth father in cold blood and fled the scene with Abigail discarded behind the bins like she was no better than garbage. By the time the police arrived, Abigail wasn't even crying any more. Like she knew that the worst had already happened, and her screams would go unheard. It took them over an hour to find her – I'd told the woman to hide her, keep her safe, not shove her so deep down into the rubbish piles that she might never be found.

I think it made me more determined to help the real victims, that case. I was never satisfied knowing the bare bones of a case and was less likely to believe they were who they said they were, or that they needed what they said they needed. So yes, I may have let myself get overly invested in my calls – even more so since Dad died and I had no one in real life who needed me. But I thought the police was the one thing I could depend on in life. The one thing that would never leave me.

You'd have thought I'd have already learned my lesson that nothing in life can be counted on after my mum upped and left when I was so young, without a backward glance. But no. It was Natalie Pierce's case twenty years later that proved to me once and for all that not everyone – or everything – is as it seems and some people don't deserve help. Maybe I have let that view impact my work, maybe losing my dad finally brought that out into the open.

But still, in Matt, Chloe and Abigail's view I'll always be the hero who talked Abigail's birth mother into storing her safely away, so she never had to see her birth father being brutally murdered. Matt and Chloe were called that night to foster Abigail until the police could trace her birth mother and decide what to do, but Natalie Pierce disappeared off the face of the earth, and so Matt and

Chloe got to keep their much-longed-for little girl. I was so distraught about the case that Amanda kept in touch with social services to reassure me that Abigail was safe. While this wasn't exactly 'by the book', it was the best thing Amanda could have done for me – and for Abigail. We formed a relationship with Chloe and Matt, and now five years on we're like one big, strange cobbled-together family.

It all worked out in the end.

–

After lunch, I collect up everyone's plates and take them into the kitchen where I fill the sink with hot, soapy water. Amanda is not the sort of mother-in-law who will flap around telling me to sit down and not lift a finger. She cooks, I clean. Or, if Nath's here, he does. The best thing you can do is marry a man who was raised by a feminist. Or better yet, marry a man raised by the feminist who lived next door to you your whole life. Nath and I never have the sort of domestic arguments I hear my team lamenting on shift. I know I'm lucky with him really, very, very lucky. He barely needs me at all. I check my phone to see if he's texted to check in and sure enough, he has. I write out a quick reply and then check my emails; despite seeing nothing new has come in I still drag my thumb down to reload the inbox, hoping to see a ping.

'Waiting to hear from Nath?' Amanda asks, appearing behind me.

'Oh, no,' I say. 'He's already texted to say how much he wishes he was here instead of stuck at work.'

Amanda shrugs. 'Well, I guess we won't be seeing him for Sunday lunch before the wedding now, will we, if he's in the middle of that big op?'

I shake my head, amazed at Amanda's power to always know exactly what's going on in the force despite being out of it for so long. I look down at my phone again, sighing, before putting it back in my pocket.

'What are you waiting for, then?' Amanda says. She stands next to me and grabs a tea towel from the side, ready to dry as I wash.

I dunk the plates in first – they're the easiest thing to wash – then hand her the first one. She dries quickly and stacks the plates to the side, nodding at me for more.

'It's just a friend,' I say. 'She's having some problems.'

'Who?'

I shake my head. 'Not someone you'd know, not from work.'

'You have friends outside of work?' Amanda says with her eyebrows furrowed.

I laugh, despite myself. 'Some! Well, one.'

'So, who's this new friend?'

'She's not new,' I say, biting my lip, wondering how to explain this situation but realising I desperately do want someone to talk about it with. 'We've been in touch for a few years.'

'What's her name?'

I laugh. 'Sally.'

'Sally,' Amanda repeats, looking at me like there's a strong chance this 'friend' is actually imaginary.

'I get her emails,' I say, as if that explains it.

'What do you mean you "get her emails"?'

I scrub at the burnt-on batter stuck to the bottom of the Yorkshire pan. 'Well, a few years ago, I got an email for Sally Jones but it wasn't meant for me. It was for another Sally Jones.'

'How did you know it wasn't meant for you?' Amanda says, confused.

'Oh, because it was from this carpet cleaning company in Weston-super-Mare that I'd never heard of following up on my enquiry.'

Amanda frowns, still confused.

'I looked at the email and could see the original enquiry was from SallyLJones@mymail.com and they'd just missed out the L and instead sent it to my email, SallyJones@mymail.com so I forwarded it on.'

'Why didn't you just email the carpet people back and say the email was wrong?'

I shrug, finally getting the last piece of batter off the tray. 'Just didn't. Anyway, this kept happening over the years. I get loads of her emails, and we just started talking.'

Amanda stops drying the Yorkshire pudding tin and looks at me, eyebrows furrowed.

'I know it sounds weird,' I say, my cheeks colouring.

There's quiet for a moment until Amanda resumes her drying motion, rubbing the blue-and-white-checked tea towel over the tin and then in circular motions to get every single droplet that might be hiding within the sunken circles.

'So, what's the problem with this other Sally Jones?' Amanda says, and for a second I'm lost, before remembering our original conversation.

'Oh, well,' I start, before realising how much I don't want to divulge more than I already have to Amanda. 'Nothing serious. Just men stuff, you know.'

Amanda sighs, like she can't believe she's already wasted this amount of time talking about something so frivolous, and the conversation ends, but I can't help thinking what if the problem with the other Sally Jones is actually quite serious after all?

CHAPTER THREE

Nath's supposed to be on a rest day today, but of course he's been called in. I'm achingly jealous, though I obviously can't tell him that because the last thing you need when you're going to work after four hours' sleep is someone lazing around at home telling you how much they wish they were you right now.

But I just don't know what to do with myself. How do people without jobs manage? I've never been someone who's dreamed of a long break; some of the women at work it's all they do, talk about their next holiday. One of them even said the only reason she had children was to get her maternity leave – and she now spends all her time telling me how work for her is a break compared to raising said children.

All the parents at work say the same. As if someone forced them to have children in the first place. I don't know if being around the constant complaints are what has made me not want children of my own, or the fact that Nath has never wanted them. The only people I know who actually seem to enjoy parenthood are Chloe and Sally. She writes about her little girl, Grace, like she's her reason for living.

I pull out my phone, checking once again for a reply from her. If Sally gets evicted, where will she and Grace go? I hate the idea of them being in trouble. Our friendship *is* odd. I know Amanda thought I was totally bonkers yesterday when I told her about Sally, but the truth is, it's been nice having a friend outside the force. My whole life has revolved around the police and it's been nice to know someone who isn't in that world; who doesn't spend their days talking about assault and murder and the worst parts of society all day long.

I pull down on my phone screen, reloading it in the hope of hearing that magic little ping that tells me a new email has come in. Nothing. Well, not nothing. The wedding photographer has messaged again asking me to confirm how many shots of the ceremony we want. I sigh, knowing I've absolutely no reason not to reply but finding myself wanting to do absolutely anything else. Then I think of Nathan's face when he comes home later today, asking me why I've not replied. He's cc'd in on all the wedding emails and up until now has been the one taking the lead – despite his hectic schedule – but he said to me last night that he was going to leave the organising to me from now on while I am off work, which is fair enough.

I tap out a quick reply to Joanie, the photographer, and add in a few smiley faces to try and make my short message more enthusiastic. It's not that I'm not excited about the wedding – I am – but since Dad died, everything has taken on a cloudy quality. Like I can't quite feel it, or it's not quite real.

Everything, it seems, except for worrying about Sally and Grace. That's a far more practical problem I can solve.

I google 'Sally Jones, Weston-super-Mare'. It's strange that in all the years we've been emailing, I've never actually googled my namesake. Everything I know about Sally, I've learned from the rogue emails that come my way, which have led onto conversations, but it's occurred to me that I might be able to find a photo of her online, something that might help allay my concerns over her wellbeing. The search loads in seconds, and one result makes my heart race.

> Local woman charged with multiple acts of animal cruelty

Shit. I click into the article, my stomach churning. What if all this time the woman I've been speaking to is a total psychopath? Eventually, the page loads and I let out a heavy sigh. It's not her. The article says the woman in question is fifty-three years old. My Sally is young, I don't know how young exactly, but Grace is only five and to be this woman, she'd have had her at forty-eight. Not impossible, but improbable. Plus, she doesn't *seem* that old. Something about the way she writes gives off a young impression – younger than me, despite being a mum.

As I read more of the article, I'm further convinced this is not my Sally. Thank god. I go back to the search bar and this time type in Sally's full address.

Nothing of use comes up. A couple of links to People Directory and other paid-for services that promise you the details of anyone you need for the low, low price of £29.99. I wish I was able to access the police database. If Sally had ever been involved in a crime – whether as a witness, victim, or even perpetrator – she'd come up straight away. But even if I were still at work, I wouldn't be able to search her. There are checks and balances against us doing this, every search must have a policing purpose and, sure, the likelihood of someone checking this is pretty low but you never know who's watching.

I know people do it. My own team has done it; I've had to haul them over the coals for it. But it's only ever been for understandable, explainable reasons. I'm not sure how I'd begin to justify looking up Sally – a woman I've never met with my same name – while on enforced leave. Plus, my inspector literally told me to stop getting myself so involved in other people's lives so I can't see how this would help my case.

As I'm scrolling through more Google results, each more useless than the last, Nathan phones to tell me he's got a few hours' spare and will pop home for lunch, so I give up the search and try and force myself into doing something more useful.

'What's up?' I ask him when he walks through the door, his face pulled into the fake smile he puts on when he's trying to pretend everything is okay. It's nothing like his real smile – the one I've seen, and loved, for the whole of my life. That smile makes his nose crinkle just a touch and his eyes soften, like he might laugh at any moment. This one, while turning his lips upwards, leaves the rest of his face blank. The last time I saw the fake smile was when our wedding caterer went bust and took our deposit with them, so naturally I'm worrying what's behind this one.

'What?' he says, hanging his keys up and shrugging off his jacket.

He looks smart today; he always does for work. He's got his best suit on, a crisp white shirt underneath with a dark blue tie. He's young to be in the position he is at work, and I know he feels the need to prove his worth every day. It's hard being the son of the former Assistant

Chief Constable – especially when Amanda was such a force to be reckoned with during her reign. Nath can be just as commanding, but he's also got a softer side that has made him one of the most popular detective inspectors on the force. Sometimes, I still don't know why he chose me; he could have anyone he wanted.

'You look shifty,' I say, walking into the kitchen as he follows. I hear him sigh behind me.

'What's for lunch?'

I spin around with a frown. 'This isn't the Fifties, Nath. You don't drop home to be served a hot lunch before you go back to your hard day's work.'

He rolls his eyes. 'Give over. I meant what shall we have.'

I click my tongue against my teeth, knowing I'm in the wrong but not knowing how to shrug off my bad mood. Nath has never, ever treated me like his second in command, let alone a housewife but there's something so utterly disempowering about being stuck at home, useless, while he goes out to work every day. I often feel snippy about the power balance between us – how I spent my entire childhood and teenage years having the world's biggest crush on him, but he didn't see me like that until five years ago. Sometimes, it feels like I'm still that silly little girl in love with the oblivious boy next door and I wish I knew what it was that caused him to finally see me as I saw him.

'There's some leftover ham in the fridge from your mum's roast yesterday,' I say.

'Great, I'll make us some sandwiches.'

He takes out the ham and slices up the loaf I remembered to buy yesterday from the bakery we like but never usually manage to get to before it closes on a

Sunday. I watch, guiltily, knowing I should be making lunch. I'm so keen to help everyone else but can't bring myself to do this. Why not? It would have been the kind thing to do – it's what he'd have done for me. I hate this side of myself – the pettiness, the desperation to never be seen as Nath's 'wife', the same way both of us never wanted to be known as the respective children of the famed Assistant Chief Constable Amanda Yates and well-loved Chief Superintendent Graham Jones. A few years ago, I thought about going for the Fast Track scheme to move out of the OCC and into being a detective. I could have done it, I'd got the right type of degree, the right kind of experience. But I couldn't stand the idea of being thought of the way I know everyone sees those detectives. It's stupid, but you don't get half the respect that the officers who come up the traditional route do. Nath said I should go in as a PC and work my way up but the idea of halving my salary and having to rely on Nath's income to pay our mortgage put me right off. He, of course, said I was being ridiculous and that's what partners did for each other, but Dad raised me to be independent – not to rely on anyone. So instead, I stayed in the OCC and tried to curtail my envy as Nath rose through the ranks and I remained on the sidelines.

'Here you go,' he says, placing my sandwich in front of me. My stomach rumbles as I realise I've not eaten anything today. I mumble my thanks and tuck in. 'You got much planned for today?' he asks cautiously and I let out a little laugh.

'Busy day,' I say. 'Busy, busy.'

'I'd gladly swap you.'

'Why, what's going on at work?' I lean forward, desperate to be included in something that matters.

'Oh, just the usual. The Chief is breathing down my neck about the budget and my sergeants are complaining they can't give me more PCs if I won't approve more overtime.'

I sigh. 'Bit weird that you've managed to come home in the middle of a shift, given you're so busy?'

He shifts in his seat, his eyes firmly on his sandwich. 'Just wanted to see how you were.'

'Why?'

He doesn't say anything.

'Has something happened at the OCC?'

He looks up, that guilty fake smile back on his annoyingly handsome face and I tut impatiently.

'For god's sake, Nath, just tell me.'

'Look, I don't know anything for sure but there are whispers they're bringing in someone else to take over your shifts.'

I swallow, the sandwich now sitting heavy and uncomfortable in my stomach. 'What did you hear? Tell me exactly what they said.'

Nath pushes his plate away and shakes his head. 'Honestly, I can't even remember but there was just some talk about it in the office. You know what people are like, they love all that crap. They stopped talking about it as soon as they realised I could hear, obviously, and I wasn't about to force them into telling me—'

'You should have!'

'Come on, Sal. You know I can't be seen to be getting involved and I'm not going to start gossiping with the call handlers.' He frowns at me, like he's explaining workplace dynamics to a child. I roll my eyes, not exactly proving myself to be above this. 'I just wanted to come and check in on you. There's nothing more sinister behind it. Alright?'

He stands up, pushing his chair out noisily behind him. 'I won't bother next time if this is the reception I'm going to get.'

'Nath,' I say, calling to him as he walks to the front door. 'I'm sorry. I'm just…'

He stops. 'I know,' he says, turning around. 'But honestly, I'm worried about you. You need to keep yourself busy. Get out and do something, don't just sit in the house and stew all day. It's not good for you.' He leans in for a kiss before grabbing his keys and telling me he'll see me later.

He's right. This isn't like me. I'm not a person who sits around doing nothing, waiting for things to happen. It was my uncharacteristic inaction that led to this situation in the first place; if I'd been more on it at work rather than letting my emotions derail me, I wouldn't be at risk of losing my job. Because that's what will happen if I'm gone for too long – if they bring in someone else who's better than me, I might never get back to work. And I need to work, I need to help people – it's what I've always done, and right now the only person who I can give that help to is Sally. Whatever the situation is with her eviction, there must be something I can do to help her. If she's too proud to come to me for help, perhaps I need to bring the help to her doorstep.

CHAPTER FOUR

'What do you mean you're going to Weston?' Nathan says as I try to slide the empty overnight bag back under the bed.

'You were the one who said I needed to get out and do something,' I reply.

'I meant sort stuff for the wedding. Go for a walk. Have a bloody spa day, I don't know.'

'When have I ever had a spa day?' I say with derision, forgetting momentarily I'm the one in the wrong here. I haven't exactly told Nathan about my planned trip, or the reason behind it.

'You're going overnight?' he asks, his eyes full of confusion as he watches me sling some pyjamas into my bag.

'No,' I say, shrugging. 'Well, probably not. No. I just want to see the sea, I'll probably be back tonight – but just in case, you know.' I gesture towards the pyjamas and continue packing.

'Sal,' he says, pulling me to sit beside him on the bed. 'What's going on? Really.'

I bite my lip. Perhaps I should tell him the truth, but the thing is, I wouldn't know where to begin. There's no reason I've not told him about Sally before now – nothing that will make any sense to him, anyway. It's just that the friendship I have with Sally is a bit weird – kooky, I guess. And I'm not really that kind of person. Nath has always known me to be sensible, straightforward – easy. He'd have told me not to share details of my life with a stranger online, that I have plenty of friends at work if I need someone to talk to. I never knew how to explain to him that I wanted something outside of the job, someone who didn't know me as former Chief Superintendent Graham Jones's daughter, or Nath's fiancée, or even the Assistant bloody Constable's daughter-in-law to be. I wanted my own identity. Talking to Sally gave me that.

'Nothing is going on. For god's sake, I just want to go to the seaside for the day! You're being ridiculous,' I say to Nathan now, hiding my embarrassment with righteous indignation.

'Have you got back to the photographer yet?' he shoots back.

'Yes, I told her the shots we wanted—'

'But what about the timings? She asked you for a timetable of the day.' He glares at me, and I rack my brain trying to think about when this request came in and if I have, in fact, ignored it as he's suggesting. 'Fuck's sake, Sal. Do you even want to get married?'

'Of course I do!'

'Yet you can't even be bothered to reply to our photographer?'

I say nothing in reply to this accusation because I have no excuse.

'Before, I just thought your haphazard approach to organising the day was because you were busy at work and after we lost your dad, I thought it was just a temporary loss of interest. But now all you do is complain how bored you are and that you've got literally nothing to do and yet you still can't even bring yourself to reply to our photographer! It makes me wonder. It really fucking does.'

He gets up from the bed and strides across the room.

'Nath,' I say. 'Don't be like that.'

He stops, his back to me as he reaches the bedroom door. 'I've got to go,' he says, his voice detached, resigned to my nonsense. 'Enjoy the seaside.'

I hear his heavy footsteps as they pound down the stairs and out of the front door, slamming it behind him. I let out a sigh and flop back in the bed. I understand his frustration; I'd be frustrated too. But I need to find Sally, I need to check she's okay. Once I've done that, everything can go back to normal. I'll reply to the many ridiculous wedding emails, I'll go back to work, we'll get married and normal life can resume.

But first, I need to find my friend.

-

As I drive down the M5 with Radio One blaring, sunglasses on, I think how mad it is that Nath and I don't do this more often. It's only two hours from ours to Weston and yet we've never gone there as a couple. I used to go with my mum and dad when I was little, before she left. I remember them packing up the car with a picnic and leaving at the crack of dawn to make the most of a sunny day. We'd spend the day on the beach and if I was lucky, they'd treat me to a donkey ride. Dad and I would drift off in the afternoon and spend hours in the arcade while Mum topped up her tan. I loved the two pence machine and Dad would laugh as I endlessly fed it our money, squealing with delight when the mound of pennies finally tipped over and collapsed onto the tray below, ready for me to collect. After Mum left, Dad and Amanda took Nath and I a couple of times and we'd wade into the sea together – trekking what felt like miles through the sticky mud to reach the shoreline. In other circumstances, I could have brought Nath with me today and we could have made a day of it. But I have to remember I'm not off for a fun trip to the seaside, like I told him. I'm going for a very specific purpose.

As I arrive in Weston, the sun goes in. It's not exactly a warm day but the sunshine on the drive down totally lifted my spirits. This has been the wettest spring on record and I'd not realised how much the weather dictated my mood until I saw the sun. Now it's gone again, the day feels less like an adventure and a bit more ridiculous. This is mad, isn't it? Driving all this way for a woman who won't even recognise me – let alone need me. I think back to my supervisor's words, *you want to help everyone, but you also have to look after yourself*, and imagine what she'd say if she could see me now. This is a stupid idea. But what if I'm right and Sally does need help but hasn't felt she could say so on email? Some things are easier to say in person, so I keep driving.

Plus, I've already planned what I'll say when I show up at Sally's door. *It's me, Sally Jones from your emails!* I'll say with a laugh and a little wave. *I was in Weston for the day and thought I'd pop by and say hello in real life.* She'll probably then invite me in for a coffee and explain why she's been so quiet and everything will be fine. I'll go back home, safe in the knowledge that nothing terrible has happened to my friend, and we can resume our email chats. Maybe she'll invite me down again, this time without me showing up on her doorstep, and we can become proper friends. How we first met will become part of a funny story once we know each other in real life, instead of this strange vaguely embarrassing online friendship that I've not told anyone about. I could even invite her to the wedding! Maybe she'd bring along Grace, she's around the same age as Abigail – they could play together. It would be lovely to see the two of them together, all dressed up for the day, while Nath and I saunter around the venue, happily married.

But still, best not to get ahead of myself. I've got to find Sally first – she might not even be at the flat after that eviction notice.

I find somewhere to park in the centre of Weston that Google tells me is only twenty minutes' walk from Sally's flat. I think I need to gear myself up to making the visit, spend a little time walking around the town so if I do see Sally, it won't be a total lie that I'm having a day out. I pay the extortionate fourteen-pound all-day parking charge and head towards the beach.

Weston is more run down than I remember. As a kid, it was a place of wonder. All bright painted signs and sunshine, but as I look around now, it feels a bit like the town that time forgot. The signs I remember are still here, but the paint is chipping off them and many of the shops on the seafront are boarded over. Without the sun, the wind has a chill to it and I pull the sleeves of my fleece down, wishing I'd brought a coat. I walk along the seafront until I reach the pier and find a stand serving coffee.

'Can I have a flat white, please?' I ask the acne-ridden teen who looks at me with boredom before turning his back on me to make my coffee.

'Three pounds seventy-five,' he says, as he hands me the hot drink.

Clearly prices aren't stuck in the Nineties, even if the rest of the town is. I tap my card to pay and head back to the beach. The coffee is so hot the cup burns my fingers, and I have to stop and put it down on the beach wall. I sit and watch people go by. A young mum pushes a pram with a coddled sleeping baby inside it, her toddler running ahead wildly as she pulls him back by his toddler reins. She looks knackered. An older couple stroll hand in hand behind her with a yappy white dog. They don't stop when he takes a steaming shit on the pavement, the smell of which is blown back to me on a gust of wind. I pick up my coffee, burnt fingers be damned, and move on.

I carry on along the seafront in the vague direction of Sally's flat. I wonder how many times she and Grace have walked up and down this path. It must be nice to bring up a child by the sea, though I've never much yearned for it myself. The raising children part, not the sea. I know Amanda is desperate for a grandchild and it is sad that we won't ever give her one but still, you should have a very good reason for having children and wanting to make someone else a grandparent is not good enough. There is nothing worse for a child than to be born to parents who don't really want them. I know that well enough and perhaps if my mum had considered her options a bit more wisely... Well, then I wouldn't exist, I suppose.

Google Maps tells me to turn left, away from the seafront and I pause, not sure if I want to follow this path after all. I look back to the beach – the sea is so far out you'd have to wade through the thick mud to get there – but still, the slow rippling waves are a comfort to watch. I take a seat on the closest bench and pull out my phone, needing to remind myself that Sally is someone worth searching for. That she isn't a stranger I barely know but a friend who might need my help. I pull up an email Sally sent to me last year in reply to something I'd received meant for her and forwarded on.

> Subject: Re: Your Free Carpet Fitting
>
> Hi Sal,
>
> Thanks for sending this one across – sorry you got it again! They keep putting bloody leaflets through the door as well. I went in a few weeks ago to see about getting some new carpet for my stairs and they won't give up. You can just delete the rest if you get any more – sorry!
>
> But yeah other than my crap carpet, all okay here. Took Grace down to the beach yesterday given we had half a day of sun – first for ages! She loves the beach. We made sandcastles then got a cone of chips while the sun set. Perfect day, really.
>
> Hope it was sunny where you are! Though not the same when you haven't got a beach to go to, I guess.
>
> Sally x

I'd replied telling her I'd not had a chance to see the sun as I had been stuck inside the bunker for the last sixteen hours, but was glad she and Grace got to go out. I re-read her message. She sounded happy, like life was easy. I had pictured her standing in her wide hallway looking up at her old staircase and thinking it could do with a refresh. She's never told me her hallway was wide, or that it had those beautiful black-and-white geometric patterned tiles you see in interior design magazines, but that's how I've always pictured it. She wrote like a person with a beautiful home.

Well, until the rent arrears notices started coming through, anyway. It made me realise I might have invented a life for Sally that didn't match the reality. But it's not like I care about any of that stuff. Our house is not something you'd see in a magazine. It's a two-bed terrace near the town centre, two up, two down. We could move somewhere bigger, but neither of us can be bothered. While I might secretly love the idea of having the wide hallway with the fancy tiles, I'm not going to spend any time trying to make it happen.

So finding out Sally's flat isn't in the nicest area of town isn't exactly a big deal, although I do find myself gripping the straps of my backpack a little tighter the further into the estate I get. Her flat must be close, she said she could see the sea…

I stop. Something isn't right. Then, I realise. Why would you need carpet for your stairs if you lived in a flat?

I frown. Perhaps it's a duplex flat. Nath and I used to live in one of those in Stratford when we first moved in together, like a mini house of sorts but inside a flat. Yes, that must be it.

A drop of rain wets my cheek, and I take it as a sign the universe needs me to get up and move on. Time to see Sally's flat for myself.

CHAPTER FIVE

The further I walk from the beach, the tattier the streets get. It's been less than ten minutes but already it's worlds away from the seafront – and that was hardly the nicest of areas, with fag butts and dog poo littering the pavements. I look down at my phone again as it tells me to keep going for another three minutes. The houses either side of me have turned to flats and straight up ahead is a row of tower blocks. We have a fair few blocks like this on our patch the further north you go in Warwickshire and, although I hate the stereotype, they are the places we get called out to most. I look at Sally's address again and wonder if she does in fact, live inside the block. Not what I imagined for her. Not at all.

As the map shows me getting closer to Sally's address, the chances of her not living in the tower block get slimmer. This doesn't fit with the life Sally has told me about over email. Have I got everything totally wrong? Assumed she's someone she isn't? I'm almost certain she said she lived in a house, not a flat. We talked about how annoying old-fashioned traditional fireplaces are to get cleaned; we joked how we're both terrible gardeners, that everything we plant dies no matter how much time (or how little) we spend tending to them. Looking around, there's no way Sally has her own garden. Not living here. But perhaps it's a shared space. I'm being snobby. She might be part of a local collective; perhaps they share an allotment. Somewhere to take Grace to learn about soil and plants...

'You lost?' a voice to my side says, making me jump. The voice belongs to a teenager in a red hoody and black joggers. She frowns at me, the lines in her forehead deeper than they should be for someone so young.

'No,' I say, as cool as I can.

'You *look* lost,' she says with a smirk.

'I'm visiting a friend,' I say and even I can hear how prim it sounds – how totally out of place I am. I never think I'm particularly posh, but sometimes I catch a certain rounded-out word and clock it.

'Here?' The teenager laughs. I tell her yes, feeling the blush on my cheek giving me away. 'What's she called? Your friend.'

I meet the girl's eye and debate whether to tell her the truth. She takes out a pack of cigarettes, Marlborough Reds, and lights one up letting the smoke blow in my face. She smiles. It makes her look even younger.

'Sally,' I say.

'Sally what?'

'Sally Jones,' I snap back before realising I've literally no obligation to tell this girl anything but there's something so commanding about her that it slips out.

She smirks. 'Sounds like a made-up name.'

'Well it's not, it's actually—' I stop myself. I was about to tell her it's also *my* name, but to explain this coincidence would be going into way too much detail with a stranger. The girl stares at me, one thick black eyebrow raised comically.

I start to walk off in the direction Google Maps tells me Sally's flat is in, and the girl walks in step with me.

'Why are you following me?'

She shrugs. 'Keeping my neighbourhood safe,' she says with a smile, before taking a long drag on her cigarette. 'You never know what people like you are up to.'

I roll my eyes but find myself smiling. She's oddly charming – even as she blows smoke in my face.

'Where do you live?' I ask after a few paces. 'Up there?' I gesture towards the tall, imposing tower block in front of us.

She shakes her head and says, 'My mum taught me never to talk to strangers,' in a voice so syrupy sweet it makes me laugh. 'Specially not weird ones that hang around on the street asking you where you live.'

I laugh despite myself as I look down at my phone, which tells me we're a few feet away from Sally's address. 'I think that's where my friend lives.'

'Sally Jones?' she says again, emphasising how totally made up she deems this name to be.

'Yes,' I say.

'You *think* that's where she lives. Right...' she replies, taking a final drag on her cigarette before flicking it into a dying bush next to us.

'It *is* where she lives,' I say back, irritated now.

'I'll bet you a tenner there's no "Sally Jones" in there,' she says.

I stare at her, incredulously. 'Okay,' I say, 'you're on.' We shake hands and I'm surprised to find hers are small and soft.

She smiles, strutting beside me. 'Wicked,' she says. 'Let's go find Sally Jones.'

–

For some reason, I'm finding it weirdly comforting to have company while looking for Sally. It makes me feel less on edge that Sally's painted a false picture of her life to me. The girl's name is Tiff, she tells me, and she's supposed to be at school, but she hates it there so she's not going any more. Maybe she's lonely. She's got this spiky attitude like everything around her is totally ridiculous, but the way she's treating our quest to find Sally like a game makes me realise how young she is. Where's her mum? The responsible thing would be to march her home and I'm having to fight my inner saviour complex to not get caught up in mothering Tiff because it won't help me find Sally.

'What number is it?' Tiff asks again as we continue to stomp up the dark stairwell. There's a drop of sweat dripping down my back and my chest is far tighter than it should be. This is what happens when you work in a bunker, I suppose. Not much chance for exercise when you do shift work. At least, that's what I tell myself. We've got a gym on site but I never, ever go. Bad enough being stuck in the bunker with my colleagues; I don't need to see them sweating in workout gear in my brief breaks.

I check my phone again, even though I've memorised the address. 'Five three four,' I say.

'Fucking hell!' Tiff says, drawing out the 'e' in hell so that it lasts far longer than it should. 'That means it's on the fifth floor! Five bloody floors!'

'You don't have to come, you know,' I say back and that, at least, shuts her up as we continue the final few sets of stairs in silence, except for my ragged breath, which makes me feel more unfit with every single step. How does Sally manage getting Grace up and down these stairs every day? My unease grows with every step we climb; I can't picture my Sally here at all.

'I want my tenner,' she mumbles, and I tut. Do I even have a tenner on me? It's not like anyone carries cash any more. Part of me hopes that I *do* have to pay out, that I've somehow got this address wrong, and my Sally is safe in her fancy house with her messy fire pit like I've always imagined. Because if my Sally does live here, then I've been wrong about her all along.

Finally, we reach the fifth floor and despite my growing unease, I turn around to share a grin with Tiff, who looks as knackered as me, if not more.

'Fucking hell,' she says again but there is a look of glee in her eyes, like we've climbed Everest and actually, it does feel a bit like that.

'Right,' I say, trying to calm myself. 'Five three four, five three four...' I wander up and down the dark corridor that's pungent with the smell of urine.

'Rank up here,' Tiff says. 'Thought our floor was bad.'

'Which floor are you on?' I ask as I count the numbers going up, 510, 512...

'First. Mum wouldn't take the flat on the ground floor. Says they get robbed way more. And she wouldn't take any higher than first 'cause she couldn't be arsed to do them stairs. Meant it took ages to get us moved but she was right, wasn't she? Cause it's a proper shithole up here.'

Tiff isn't wrong. The corridor is dingy with hardly any natural light coming in and where it does, so much dust dances around in the sunlight that you almost wish there was no light at all. It feels so dirty. The Sally I've been speaking to would never live here. *Never.*

Nath comes to places like this all the time and though I've watched a lot of body cam footage entering and exiting these types of flats, I'm usually safely observing via screens from the bunker so it's not my natural habitat. I'm trying so hard not to judge Sally, to wonder why she's lied – or at least, misled me – about the type of place she lives in, the kind of life she lives. But with every step further into the piss-smelling hallway, it gets a little harder. This was a bad idea. I've no idea what I'm walking into and now I've dragged a teenager into this with me too. We should turn back. This has gone from feeling like a silly adventure to something stupid – no, worse than stupid – dangerous.

'Here it is!' Tiff shouts from up ahead where she's run in front of me, and I know it's too late to turn back now. She's already banging on the door with her fist, an excited sheen of sweat on her forehead. 'Open up!' she shouts. I run and drag her away from the door.

'Stop it!' I hiss, sounding like an out-of-control teacher and Tiff responds appropriately by rolling her eyes. 'You've no idea who's in there.'

I step back towards the door and knock more politely before saying, 'Sally? It's me, Sally.'

'Your name's Sally?' Tiff says from behind me with a laugh.

I ignore her and knock again. 'Sally? Are you in there?'

'You're Sally and your friend is called Sally. You're telling me this isn't made up? Hand over that tenner now, *Sally*!'

'Will you shut up for a second?' I snap at her. 'I'm trying to listen.' I press my ear to the door, hoping to hear Sally's voice as if everything is fine, but there's nothing. I step back, ready to leave but Tiff starts bashing the door.

'Oi! Sally, you've got Sally out here!' She laughs like this is the funniest thing she's ever heard and gives the door another powerful bash.

'Tiffany!' I shriek as the door flies open. 'You've broken the bloody door—'

'Wasn't me,' she says jumping back and pointing. 'Look.' She points at the lock on the door and then I realise she's right, it wasn't her fault.

Someone has been here before us, and it looks like they weren't too happy.

CHAPTER SIX

I tentatively push open the violently bashed-in door.

'Sally?' I say again, but I'm met with silence as the door opens to reveal her flat.

It's chaos.

Utter chaos.

'Woah,' Tiff says, stepping in behind me.

My eyes scan the small room. Stuff is everywhere. The orange sofa cushions have been pulled off and are scattered around the room. A cheap, MDF coffee table is overturned on the floor, pieces of a smashed mug beside it. Coffee has stained the dark grey carpet, but how long ago it's hard to tell because the whole carpet is filthy. I briefly wonder why Sally ignored all those special offers the carpet company sent to her (or, more accurately, to me) before realising that a dirty floor is the least worrying thing about this scene.

'We should leave,' I tell Tiff, already backing out of the flat.

'What?' she says incredulously. 'What about your mate? She could still be in there!'

I shake my head. 'This isn't right. This is a break-in. We need to call the police.' At the mention of police, Tiff's face hardens.

'Nah, come on, we've got to at least check she ain't still here first.' Before I can say anything, Tiff's walked across the tiny open-plan living space and into the only other room in the flat. I follow her, stepping carefully over the glass and general mayhem on the way, ignoring everything I know about crime scenes and how not to mess them up because she's right, what if Sally is hiding in here, hurt and alone?

'Worse in here,' Tiff says when I find her in the bedroom. She's right; it's even more of a state. The mattress – no bed frame, I note – has been torn through with a knife so that the springs inside poke out at vicious angles. A yellowing pillow has survived unscathed but the sight of it makes me worry even more about Sally. Why does she only have one sad, dirty pillow and a mattress on the filthy floor instead of a bed frame? Where does Grace sleep? More to the point, what the hell happened here?

'Your mate ain't here. But it looks like she has seriously pissed someone off,' Tiff says, knowingly, as she lifts up the uncovered duvet that's also been slashed, its insides spilling out.

'I don't understand,' I say.

'Probably owed someone money,' Tiff shrugs. 'Not you, is it?'

'What?'

She smirks. 'You could be a dealer, I 'spose.' She cocks her head to study me further. 'Some right poshos do it, though they don't usually come to places like this in person.'

I frown and shake my head, shocked someone as young as Tiff could be so worldly already. But I shouldn't be surprised. I read about people like her all the time, take their calls, listen to the kids' voices in the background of the chaos. But still, it's different in real life.

'I'm not a *dealer*,' I say, instantly regretting my use of air quotes, which crack Tiff up.

'But is she really your friend?' Tiff asks, once she's composed herself. 'This Sally?'

I sigh and it's enough to show Tiff things aren't quite as I've portrayed them. 'She is my friend, but… I've not met her in real life before.'

She raises her eyebrows, grinning. 'Tinder?'

'What?' I say scrunching up my nose. 'No, I'm engaged.' I hold up my finger, flashing my ring, instantly regretting the show of wealth in such dire circumstances.

She shrugs. 'That don't stop most people.'

'To a man,' I add, and she rolls her eyes. 'I've spoken to Sally via email for a couple of years. I came today because I found out she was being threatened with eviction and I was worried about her.'

Tiff considers this, looking around the wreckage of the room again. 'Landlord won't have done this, even if she didn't pay. They're shits but they're not criminals.'

'No,' I say. 'I think this is something else altogether.'

I walk back out through the bedroom into the living room, looking for any clues about what might have happened here. What is Sally mixed up in? The scene is violent. Whoever did this wasn't looking for something to steal – the wreckage feels personal. I lean down to press my hand against the discoloured carpet next to the broken mug.

'It's not wet any more,' I say to Tiff, who's followed me. She tilts her head to the side like she doesn't understand. 'So either this stain is old, or the mug was broken long enough ago that whatever spilled out of it has had time to dry.'

'Alright Columbo,' she says.

'You're too young to know who Columbo is,' I retort.

'So are you,' she shrugs.

I walk around the room again, looking out of the dirty windows. It's high up here; pigeon shit smears the view but the sea is right there. Sally's told me about this before, how her favourite thing to do is sit with a cup of coffee in the morning and watch the tide come in. It's a million miles away from the scene she'd conjured up for me, but not a complete lie.

'It's not messed up in here,' Tiff shouts from across the other side of the flat where she's snuck behind a door I didn't notice before.

I follow her voice into the tiny, mould-ridden bathroom. She's right, it doesn't look like it's been touched though that's hardly saying much. It's a horrible little room, with no window and a bathroom suite that looks easily more than twenty years old.

'I can't imagine how anyone could manage in a bathroom this small with a kid,' I say.

'It's alright,' Tiff says back. 'There's three of us in ours and it ain't that bad.'

I say nothing because, of course, I hadn't considered that Tiff's flat is probably the mirror image of this one.

'Not sure she's got a kid in here though,' she adds.

'Why do you say that?'

She turns to me like I'm an idiot. 'Well, where's all the kid's stuff?' She gestures to the bathroom around us. The Head & Shoulders two-in-one bottle of shampoo and conditioner, the single bristled toothbrush on the sink, the half-empty luminous green shower gel.

'Our bathroom's full of kid shit,' she carries on. 'You can barely get in the shower around it. Pisses me off.' She kicks at the skirting board and looks down.

'Sally has a five-year-old,' I say. 'Grace.'

Tiff shrugs and walks out, no longer interested.

My phone vibrates in my pocket, Nath's name on the screen. Shit. How am I ever going to explain what I'm doing here? I'll have to tell him about the break in, especially once I report it. Though, the chances of him finding out are quite small. It's a different force down here, so how likely is it that they'd contact Warwickshire to let them know I was involved in a crime? Well, I suppose it depends how serious this case gets. If we can't find Sally, and her flat's been wrecked like this, I imagine it could get very serious indeed. That's the last thing I need. How will it look while I'm supposed to be off 'recuperating' that I've got myself involved in a crime scene? Walked all over it with my stupid shoes and contaminated the evidence? And brought a minor into it?

As I walk back into the living room, Nath rings off. Good. I don't have time to deal with him right now – not until I've found out what the hell has happened to Sally and Grace.

'Come on,' Tiff says, beckoning me to leave the flat. 'Giving me the creeps in here. She's not here, is she?'

She walks out of the flat as I spot a notebook hidden just under the sofa by the overturned coffee table. I pick it up, glancing only briefly at the flower-patterned cover, then shove it in my backpack to inspect properly later.

—

'Before we call the police,' I say to Tiff in the corridor, 'we should talk to some of her neighbours.'

The truth is, I'm not ready to call this in. Once I do, I'll have to step back – leave it up to the local force to help Sally out of whatever mess she's got herself into and I won't be able to do anything about it.

'No one's going to talk to you,' Tiff says bluntly.

'Why not?'

''Cause you look like a pig.'

I pause. What would Tiff say if I told her I am a 'pig'?

'I'll talk to them,' she says with a grin, and she marches off before I have a chance to consider what a bad idea this all is.

I watch, helplessly, as she starts banging on the doors in the corridor. None of them open up and I don't blame them given the way Tiff is going on.

'For god's sake,' I say, catching up to her and stopping her fist before it bashes on another door. 'Give it a rest.' I nudge her out of the way and calmly, politely knock on the door in front of us.

Despite Tiff's muttering under her breath that no one will even hear my 'posho knock', after a few moments the door cautiously opens and a woman around my age peeks out.

'Yes?' she says, keeping the latch on so that only about a quarter of her face is visible.

'Hi,' I say, putting on my best 'professional and calm' voice that comes naturally after so many years of dealing with crises. 'Do you know the woman at five three four?'

She eyes me suspiciously then looks at Tiff standing behind me. The door closes in our faces, but then reopens with the latch on, revealing the chubby baby attached to her hip. She jiggles it up and down and says, 'Why?'

'I'm a friend—' I start to say but Tiff interrupts.

'I live 'ere,' she says. 'First floor.'

'Yes,' she says. 'I recognise you.' The woman studies Tiff and turns her gaze back to me.

'The flat's been broken into,' I say. 'My friend's flat. Number five three four.'

She takes a step back. 'I don't know anything about that, sorry.' She goes to close the door but I hold my hand up.

'Please,' I say. 'Do you know her? The woman who lives there?'

The woman sighs as the baby starts to gurgle loudly. 'It's time for his feed,' she says. 'You'll have to come in.'

We follow her into the flat, which is the same size as Sally's but feels totally different. It's decorated in bright, welcoming colours and is – despite the baby paraphernalia everywhere – neat as a pin. The woman takes a seat on the yellow sofa and whips out her boob, quick as a flash, before the baby greedily sucks at it. Tiff turns her nose up beside me and I elbow her in the ribs to tell her to behave.

'I don't know her well,' the woman says. 'I only moved in last year.' She keeps her eyes on her son as Tiff does everything she can to avoid looking directly at the woman's boobs, instead inspecting every single book on the shelf in the corner.

'When did you last see her?' I ask.

The woman shrugs, which momentarily disrupts the baby's feed but he's quick to latch back on. 'Last week, maybe?'

'Did you know the flat had been broken into?' I ask.

She shakes her head. 'I don't want to get involved. It's best to mind your own business here,' she says, raising her eyebrows, and I nod, understanding.

'Please,' I say, leaning forward. 'If you know anything, it would really help us.'

The woman pauses, as if considering this but then looks down at her son, shaking her head. 'I don't know anything,' she says, and I slump back.

'Is her name Sally, though?' Tiff says, still without looking at the woman. 'Is that her actual name?'

The woman thinks for a second, as if debating whether this is safe information to share and my pulse races. If she says yes, then this really is the home of my friend, and that friend has been lying to me for years. If she says no, this is all some big mistake that I can walk away from.

'Yes, I think it is. Something like that.'

I let out a heavy sigh. This means I can't simply wipe my hands of this and pretend it never happened, even if I wanted to. My friend really is in trouble, and I can't keep pretending otherwise.

Tiff mutters 'fuck's sake' under her breath, which again makes the woman give her a curious glance.

'Ignore her,' I say, knowing Tiff is mourning the loss of her tenner. 'I'm really worried about her, about Sally. She's not been in touch for a little while and now I see the flat like this... I'm worried something's happened to her and her daughter—'

'Daughter?' The woman says, propping her baby up now he's finished feeding, and leaning him against her shoulder as she gently pats his back.

'Grace,' I say.

The woman shakes her head, her eyebrows furrowed. 'Never seen a daughter.'

'Ha!' Tiff says, coming to sit next to me on the small sofa once the woman's boob is securely out of sight. 'Told you.'

'I don't understand...' I say. 'She told me she had a daughter. You're sure you've never seen a girl with her? She's five I think, maybe six...'

The woman shakes her head. 'I could be wrong,' she says. 'But I'm pretty sure I've only ever seen her on her own. She's not the chatty type, keeps to herself.'

The Sally I know is very chatty. We must have exchanged hundreds of emails over the years, chatting away like we're best friends. But then again, most people would describe me as keeping to myself and yet I've been just as chatty as she has over email. Perhaps like me, she finds it easier to talk to people when they're not facing her in real life. Or, perhaps she's not the person I thought she was at all. A shiver comes over me at the idea that Sally has been deceiving me for years.

'And you didn't see anything happening at her flat this week?' I ask, trying to ignore the worrying thoughts in my head.

She shakes her head.

'No loud noises or anything like that?'

She keeps shaking.

After a moment of awkward silence, she says, 'Look, all I know is that your friend has been back since the flat was wrecked. Then, she left again. So I know whatever happened in there, she survived it. Does that make you feel better?'

I nod, telling her thank you, though the idea of Sally 'surviving' an incident at her flat is hardly as reassuring as this woman thinks it should be.

'I'm not saying anything else,' she says, taking the baby off her shoulder and facing him to us as she bounces him on her knee. 'I just don't want to get involved.'

She stands up and I know it's time for us to leave so I nudge Tiff to get up and we walk out of the flat, thanking her as we go. She saw something, I know she did. But I can tell when someone is frightened, and I don't blame her for prioritising the safety of her baby over helping two strangers.

Back out in the corridor, I stare at Sally's wrecked front door, thick dread sitting in my stomach. Something is seriously wrong here. Whether Sally has been lying to me or not, I can't just turn my back on her. But first, I'll have to find her. Where the hell has this woman gone?

CHAPTER SEVEN

'You'll text me if you hear anything?' I say to Tiff for the second time, and she shakes her head.

'I'll *WhatsApp* you,' she says as if I've asked her to correspond by quill-written letter.

'You know I've got to report this, right?' I say and she huffs, blowing a piece of errant hair out of her eyes.

We're back on the street now and the sunlight is a stark contrast to the dinginess of the flats. The wind whips at my face but I'm glad of the air, despite the cigarette smoke billowing from the groups of lads stood nearby. She doesn't reply either way, so I say goodbye and walk off back towards the seafront.

As I walk, I consider my options. I should have already phoned the police. But the moment I do, I know I'll have to back off completely. I can't explain it, but I feel like Sally needs me to help her. Not just anyone, but *me*. Plus, if I report the break-in, the police are going to ask how I know Sally and what my involvement is. I'll have to explain to Nath why I came down here, that I lied to him. That I can probably deal with, as well as explaining things to the police, but how can I give up the chance I have to help Sally?

My pace picks up as I think this through. I don't have any other option. This isn't about me; I have to call it in. Yes, my involvement in this is going to make me look even more emotionally unstable than everyone thinks I am already, and I'll probably never find out what happened in that flat, but what else can I do? This is about Sally and Grace – about keeping them safe – and the police are the people best placed to do that.

As I reach the seafront, I walk past a group of young kids in school uniform with their mums trailing behind them. I search their faces as if I'd even recognise Sally if I saw her. It's ridiculous that I have no idea what she looks like. I've searched for her on social media, obviously, but there are so many Sally Joneses I've never been able to find which one is her; there's never been anyone who's stood out to me as who I imagine her to be. But then I've clearly had Sally all wrong, haven't I?

I consider the fact that Grace has never been seen by Sally's neighbour. But how reliable is the neighbour really? She wouldn't even tell me what she'd seen going on at Sally's flat, plus she only moved in a year ago. But a year is a long time. I see our neighbours coming and going every day. Is there any chance one of them could have a daughter I'd never even seen? No. The idea is ridiculous.

I sit back on the wall by the coffee shack I visited earlier and root through my bag to find my wallet.

Instead, my fingers find Sally's notebook.

With the shock of everything, I'd forgotten I'd taken it. I flip open the first page but it's empty. I then realise this isn't the first page at all, there are pages missing – ripped out. I flick through the rest of the pages, frustrated as the blank white paper stares back at me. Then, something falls out. Jumping off the wall, I grab it before the wind takes it.

It's a black-and-white newspaper article and as I unfold it, a prickle of unease unfurls across the back of my neck.

I recognise the article well; I was so proud when it came out. I stare at the photo of myself from five years ago under the headline '*Chief's Commendation for brave 999 call handler*'. In it, I'm stood next to the Chief Constable, holding my award, my face beaming and cheeks red. The local press had come out that night to the award ceremony and I couldn't believe it when they wanted to take my picture. I'd never been in the papers before, or since.

> Warwickshire Police call handler Sally Jones was awarded a Chief's Commendation in last week's award ceremony at Leek Wotton for her brave handling of a harrowing 999 call that came in last year. Her swift actions saved the life of a helpless baby after she was left at the scene of the murder of known criminal Shane Blackwell in December. Though Ward's killer has yet to be found, Chief Constable Andy Argent said that Jones' decisive actions that night made all the difference to ensuring the wellbeing of the baby. Natalie Pierce, 20, of Lewis Road, Coventry, is still wanted in connection with the incident and police are appealing for any information on her whereabouts. Should you have information, you are urged to call 101 quoting crime reference number 63201219 or report it anonymously to Crime Stoppers with reference 10993.

Why the hell does Sally have this article about me in her flat?

It doesn't make any sense. I flip it over, as if something on the back might give me a clue but the other side is nothing of note – an advertisement for a funeral home. This is no coincidence; Sally cut this article out for a reason. My unease grows the longer I look at it; I need to walk, to move. I pick the notebook up to put the article back inside its pages, but my hands are so shaky I drop them both on the floor.

'No!' I shout as a gust of wind picks the article up and whisks it off in the direction of the sea.

I run, trying to chase it across the beach but it's gone.

'Fuck's sake,' I scream into the wind as a group of teenagers laugh at my failed efforts to catch the paper.

Embarrassed and frustrated, I go back to the wall to collect myself.

The article is five years old. Printed before I made contact with Sally, I'm sure of it. I get out my phone and scroll back to that first ever misdirected email. Sure enough, it shows that we didn't communicate until months after the article was out.

Is there a possibility Sally already knew who I was? I suppose she could have found the newspaper article after we started talking and kept it for some odd reason. Or perhaps she cut it out when she saw her own name in the paper, and it had nothing to do with me? Either way, it's seriously strange behaviour. There's no explanation that makes any sense but I'm finding it hard to believe this is some bizarre coincidence.

My phone starts ringing in my hand, Nath's name on the screen again. If I don't answer, he'll only call back.

'Hello?'

'Hi,' I say, the wind forcing me to put one finger in my ear to hear him.

'Where are you?'

'On the beach,' I reply, standing up to keep walking.

'I can barely hear you,' he says, his voice spiky and I can tell he's annoyed. 'When are you coming home?'

'Soon,' I say back.

'I've got to do a double shift,' he says – or at least that's what I think he says, the wind is so loud I can barely make out his words. 'I'll be home around midnight.'

'Okay,' I tell him, turning away from the seafront back towards where I parked my car this morning. As I do, a group of young lads on a stag-do walk by chanting 'lads on tour!' at the top of their voices.

'Sounds like Weston hasn't changed,' Nath says, and I can hear the smile in his voice now the wind has died down. 'How *is* the seaside?'

'Windy,' I manage with a laugh.

'Have you had a good day?'

I sigh and let the lie slip out. 'Yeah, it's been nice. Always good to see the sea…'

'Good,' he mutters. 'I've had a shit day here…' I wait for him to go on, knowing if I push for details he'll clam up but if I give him room it usually comes spilling out. 'Half my officers got called out to an incident in Nuneaton where they found a body. Woman had been left in her flat for days. Looks like foul play so I'll no doubt lose half my team for the next week or so. Bloody typical. Last thing I need right now.'

I murmur my sympathy. 'Was she known to us?' I ask, switching into professional mode.

'Yeah,' he replies, but I can already tell he's switched off from the conversation as if someone else has walked into the room and taken his attention. 'The local team were sent out a few weeks ago after a domestic,' he says. 'Look, I've got to go—'

'Wait, Nath. I love you,' I say, the words overly sincere and unexpected.

He doesn't say anything for a moment, and I check my screen to see if he's still there. When I put the phone back to my ear, he's telling me he's got to go, but he loves me too. We hang up as I reach my car.

I google 'body found Nuneaton' and quickly find the force's press release about it, sparse and boring as usual with only the bare details. *The body of a woman found alone in her flat.* How can someone be dead for days and no one notice? But so many people live like that. Does Sally? Does she have anyone to care about her except me? The life she'd told me about made me believe she was surrounded by caring family and friends but what I saw today tells me otherwise. What if I'm all she has?

I trust in the police to investigate the break-in fully, of course I do. But, if Sally doesn't have any family pressing for details about her going missing, will they really prioritise it? Thousands of people go missing every month in the UK and so many of them never come back, despite police involvement. So often, it's the friends and family of the lucky ones who find information that brings the person back.

Sally needs someone who cares about her. I might be the only person that applies to, so even if it means putting myself in a tricky position I'm not going to call this in yet. I need to understand my connection with her before anything else happens. Her neighbour said she'd been back to the flat after it was wrecked so it's not like she's been abducted or anything. She obviously saw what had happened and has gone somewhere to lie low.

I walk back to my car, my hood pulled up to protect me from the wind that howls through my ears. By the time I reach the car, my cheeks feel numb and the second I get in, I switch the heating up high.

As I pull out onto the road, the sun is setting in my rear-view mirror. I'll drive home before I make any decisions. I can't think straight down here with the wind whipping everything up like this.

CHAPTER EIGHT

The drive home took forever. I was stupid to set off at peak time when every single person in the entire South West seemed to be driving up North at once. But three hours later, here I am walking back through our front door. Nath isn't due home for hours. He might not come home at all; when a shift goes on too long, he sometimes prefers to stay at his mum's who lives nearer to HQ than we do. She'll be pleased if he does; Amanda loves nothing more than having her son home for the night – even if he arrives at four a.m. and leaves by eight. She's used to that life, after all.

I've still not decided what I should do next. But first, a cup of tea. I hear Dad's voice in my head as I switch on the kettle. *Nothing was ever made worse by having a cup of tea first*. He was always telling me to slow down, to take my time, that the world would still be there when I was ready. But I never had his calm energy. Perhaps I would have been an easier-going kid if I'd had my mum around instead of her up and leaving me when I was so young. That kind of thing changes you forever. But still, I had Dad and Amanda and Nathan, it's all I've ever needed.

Once the kettle has boiled, I put a timer on my phone for four minutes to let the tea brew. It's the only way I ever make a decent cup of tea – otherwise I leave it in for a few seconds and get fed up of waiting. The countdown gives me the structure I need to get it right. As I wait, I clear up the bits Nath has left out on the side from this morning – a plate with toast crumbs, the dregs of a cup of coffee. He's usually the tidier one but he's pretty knackered with this latest case. The timer goes off and I finish making my tea then take the hot mug over to the kitchen table where I get out my phone and debate once again whether to phone the police.

Shit.

Sally's emailed me.

Her name on the screen makes my stomach curl – thank god she's okay. I click to open the email and wait impatiently as the message slowly loads.

'Come on!' I shout at the phone in annoyance.

The message comes up. It's short, a few words only.

> STOP LOOKING FOR ME. YOU'LL ONLY MAKE THINGS WORSE. IF YOU CALL THE POLICE, EVERYTHING WILL BE OVER. JUST STOP.

I read it again, trying to make sense of the message. She knows I've been to her flat. How can she possibly know that? Could she have been hiding in amongst the debris? But there's no way; Tiff and I looked everywhere. Was she watching from somewhere outside? Hidden in plain sight? I could have walked straight past her and had no idea. How do you find someone if you don't even know what they look like?

This is bad. I've got myself into something I've no place being involved with. Why does she not want me to call the police? She must be on the run from someone, or something. Drugs? It must be drugs. It's always drugs. But the woman I've emailed over the last few years doesn't seem like the type to be mixed up in… I stop myself. This is exactly my problem. It's so easy to believe there's one type of bad guy and the rest of us are good but surely, I should have learned by now that things aren't always so clear cut. I've made that mistake before and I'm not going to make it again with Sally.

My phone vibrates in my hand as I clutch it, an unknown number calling. Is it her?

'Hello?' I pant down the phone.

'Hi, Sal.' I let out a sigh at Amanda's familiar voice.

'Why are you calling me from an unknown number?'

'Oh, I don't know,' she says, huffily. 'I changed my contract and now everything is different. Quite irritating, actually.'

'Right,' I say.

'Are you okay? You sound… off,' she says, her voice slipping into detective mode.

'I'm fine, just tired.'

'Yes, Nath said you'd had a trip down to Weston?' she says, like it's a question, although she doesn't leave me time to answer. 'Are you back home now?'

'Yeah, got in not long ago.'

'Long way to go for the day?' she says again as if it's a question but it's not, so I don't say anything. 'What's in Weston?'

'A beach… arcade, you know the usual.'

'Yes, Sal, I know what's in Weston, we used to take you there when you were kids,' she says impatiently. 'I meant what's there for you?'

I sigh, picking up my now lukewarm cup of tea and walking out of the kitchen into the living room where I turn on the corner lamp. 'I just wanted to get away for the day, see the sea.'

'Have you had any news about your work issue?' I notice the phrasing 'work issue' like it's a little problem that will go away on its own if we don't look at it directly.

'No,' I say. 'Why, have you?'

She tells me she hasn't, and I sink down into the sofa.

'Nath said he heard they're thinking of bringing someone else in to take over my shifts,' I say.

'Nonsense. I won't let that happen, Sal. You just concentrate on getting yourself some rest so you can go back the moment they call, okay?'

I tell her I will, biting my lip as I think about how distinctly un-restful my day has been.

'Bit rough, Weston, isn't it?' she says, changing the subject back.

'Yeah, not the nicest place. But it has some charm.'

'I know Nick West the ACC down there, he has some stories let me tell you,' Amanda says, and I'm reminded that, of course, she would know the Assistant Chief Constable of Avon & Somerset Police. Amanda knows everyone. 'We trained together,' she goes on. 'He always hoped to end up in a big city but says Weston keeps him busy enough with all the drugs and the poverty. Not somewhere I'd go for a day out nowadays,' she says, not unkindly but in her normal matter-of-fact manner.

'Dad was always happy to take us down before,' I say.

'You were kids then,' she says. 'You'd spend all your time on the beach, and you didn't exactly notice the less desirable folk wondering around...' She has a point; my experience of the seaside town today was nothing like the fun, colourful place I remember as a child. 'Anyway, did you enjoy it?' Amanda asks.

'Enjoy it?'

'Your trip to "see the sea". Did you enjoy it?' Amanda asks.

I pause before answering. 'No,' I say before I can stop myself. 'Like you said, it wasn't as I remembered. Quite depressing, the reality of it.'

Amanda doesn't say anything for a minute, and I stop myself from filling the silence by telling her how embarrassed and hurt I feel that someone I thought was a friend has played make-believe with me for years and now wants nothing to do with me. Amanda has this way of pulling me in, making me want to bare all of my secrets for her approval. Nath sometimes says he knows I'd never leave him because I wouldn't want to lose his mum, and I know it's a joke but it's a little too close to the truth for me to laugh sometimes.

Could I confide in Amanda about Sally? She would worry about me, but she'd also be able to help. But first, she'd be pretty furious that I've been so reckless in the first place by not calling it in. When it comes to work matters, she's so squeaky clean that you wouldn't even admit to using the office printer for non-work-related reasons in front of her. Can I tell her what I've got myself caught up in and expect anything except a stern ticking off? But she doesn't work for the force any more, so perhaps she'll be less rigid now. She loves me; she'll want to help.

'Sally? Are you there?' she says down the phone and I realise I've not been listening.

'Yes, sorry, got distracted...'

'Just make sure you're looking after yourself, I said.'

'What do you mean?'

She doesn't say anything for a moment and the beat of my heart races in my chest. 'Driving all that way on your own,' she eventually says. 'You want to be careful with that car of yours. Nath says you could do with an upgrade.'

I let out a breath and tell her I will. After a few minutes of small talk about the possibility of upgrading my car and how Amanda's getting hounded by her sister for more details about our wedding day so she can plan her 'outfits' appropriately, the chance to bring up what's happening with Sally is gone. Eventually, we ring off and I sit in the living room in silence. Thinking.

After a while, I go upstairs and find my laptop in our bedroom. Sally's email warning me off has thrown me completely. Can I ignore her warning and phone the police? It feels reckless and, after today, I want to get some control back. Before I do anything, I need to work out what I know about this woman that I've been talking to for years, who I'm now pretty sure has played me for an absolute fool. On my laptop, I go to our emails and put them in date order from the first time she contacted me to today's last, threatening email.

I start noting down the facts I have about Sally Jones from the misdirected emails of hers I've received. I have her doctor's address, her dentist, the name of her landlord, where she goes for her eye test, the name of a pub she once worked at, and that she wears a size ten and once shopped at New Look.

Looking at it, it's not a lot. But perhaps enough to start with.

I start with the pub – they once emailed her to confirm they wouldn't need her in for a shift over a bank holiday weekend. I forwarded the email on, and she replied saying thank god, she could now spend the free weekend in the sun on the beach with Grace instead. Reading it again now, I frown. Why did I never question how she managed to work in a pub as a single mother of a young daughter? Who was looking after Grace while she was behind the bar? But then, perhaps her mum lives nearby or a sister. It doesn't mean Grace doesn't exist. She never mentioned any family nearby, sure, but why would she? The subject of my parents never came up either. It's embarrassing when I think of it, that I'd ever thought of Sally as a friend when we know nothing real about each other.

Until recently, I don't think I even mentioned Nathan. I only brought it up to tell her we wouldn't share a name any more when I got married. I go back to that email; she replied straight away.

> From: sallyljones@mymail.com
> To: sallyjones@mymail.com
> Subject: Re: Your Eye Test is Due
>
> Thanks for sending over the eye test confirmation. I hate going and probably won't – they still send the reminders every year though! I'll do Grace's but not my own.
> I can't believe we won't share the same name any more. Will you keep your email though? I'd miss our chats!

> From: sallyjones@mymail.com
> To: sallyljones@mymail.com
> Subject: Re: Your Eye Test is Due
>
> God, tell me about it. Literally every annual appointment I'm supposed to do is out of date. Dentist, smear, eye test. Who has the time to keep up with it all? I've no idea how you manage with a child to look after – I can't even look after myself.

> Yes – keeping the same email! I think it'll take me a while to transition to a new name. Still, most people at work already call me 'Mrs Yates' as we're both in the police and it is, unfortunately, still a bit of an old boys' club so most of the senior officers know me as 'Yatesy's Missus' which I don't love, I have to admit. I started working there before we got together so it's not like I'm there because he is! But yes, very soon I'll be Mrs Sally Yates. Not sure I'll ever get used to it.
> Sally x

She didn't respond to that one a couple of weeks ago, but then I didn't exactly ask her any questions. The next few emails aren't responded to either, and it becomes clear that her silence began after my email saying I was getting married. Is that a coincidence? Why would she care if I was getting married? She knew I was in a relationship, we never talked about it in detail, but I'd referenced my 'partner' or 'other half' here and there. Or perhaps it was my mention of us both being in the police that scared her off? I scan back through old emails, seeing if I'd mentioned it explicitly before.

None of this makes sense. The wrecked flat, the lack of any evidence that Grace exists, the newspaper article about me in her flat. But two things are clear. Sally has been lying to me, and something – or someone – has scared her so much that she's had to disappear. The question is what – and whether it has anything to do with me.

The sound of the front door going breaks my thoughts and I hear Nath calling up the stairs. I slam the laptop shut and put it on the side. For now, Sally and everything to do with this mess, is going to have to wait.

CHAPTER NINE

The next morning, Nath is back on shift again and I'm irritable after a terrible night's sleep thinking about Sally and just how little I really know about her. Whatever happens next, I need to take back some control. Unless I want to hand this over to the police and step away completely, I need to do some digging of my own. There's something going on here that involves me, and I can't simply walk away from it and sit at home twiddling my thumbs, waiting to get back to work, waiting to get married.

Sally has left a trail for me to follow, even if she doesn't realise it. The most obvious place to start is with the landlord who sent her the eviction notice that kicked this all off in the first place. But how am I going to get them to talk to me? Maybe it's better to just be honest – tell them I'm a concerned friend who hasn't heard from Sally for a while. Before I have time to talk myself out of it, I google the lettings agency and click 'call' on their number.

'Layton and Harding Lettings, Holly speaking, how may I help you?'

'Hi,' I say, trying my best to sound like a concerned friend would. 'My name's Sally Jones—'

'Sally?' she cuts in. 'Nice to hear from you, finally. Hold on, I'm going to get Alex for you.' Before I can explain I'm not *that* Sally Jones, she puts me on hold, and I sit and listen to the obnoxiously cheery hold music.

'Sally Jones,' a male voice comes on the line. 'Where the fuck have you been?'

'Oh, I...'

'I told you that if you didn't sort your shit out I'd have to get serious, didn't I?' The words are threatening but his voice is teasing, like this is all a big game. 'You ready to play ball then, eh?'

'The rent... I—'

'I know you haven't got the money,' he says, breathing heavily down the line like he's cupping the phone to his mouth, away from the rest of the office. 'But there are other ways you can pay.'

I look at the phone in horror. Who the fuck is this man?

'Meet me tomorrow,' he says.

'Where?'

He tuts. 'You're a prick tease. I'll come to the flat. Got the keys to let you back in, haven't I?' he says with a laugh. 'They aren't changing the locks until the weekend so you're in luck. See you there.' He hangs up and I'm left speechless.

I google 'Alex' and 'Layton and Harding' and there he is. He's a youngish guy – probably late twenties – in a cheap blue suit with a cat-like smile. Alex Layton. Family business, then. I look at the other listings on the website and they're all cheap. Bedsits and flats – the kind of estate agent that specialises in shitholes. Poor Sally.

From what he said, I can make an educated guess that he's been demanding sex in exchange for her rent payments and it sounds like she stopped 'paying up'. Could it have been him that wrecked her flat? But no, it didn't sound like he knows the flat is wrecked given he wants to meet her there. Plus, why would he wreck one of his own properties? He'd have to pay to sort it out. Their website proudly boats their 'LOW DEPOSITS, NO REFERENCES' claim – so it's not like he'd be able to claim the money back from her deposit.

I google 'Alex Layton' and 'charge' to see if anything comes up about him being charged with any crimes, but nothing does. It's amazing – and scary – how often you can find people's criminal charges by doing this, even with the less slimy-looking characters.

I email Sally. If this is what she's scared of, I know I can help her.

> I know what Alex Layton is doing to you. I can help. Please, Sally. Where are you? Is Grace with you? Let me help. I promise, I can help.

I listen to the message whoosh out into the ether and imagine her reading it from somewhere. Is she somewhere safe? I hope so.

'Sally?' Nathan's voice calls up the stairs and I slam the laptop lid shut.

'Up here,' I shout back, trying to keep my voice normal, and listen as he runs up the stairs.

'You're up.' He smiles widely as he opens the door.

'I thought you'd left?'

'I did, but then got told they need me on a double shift, so I need to grab some clothes for later. Didn't even have time to shower this morning and I've already got coffee down myself.'

He goes into the wardrobe and picks out another white shirt and a pair of dark trousers. What if he'd popped back a few minutes earlier and heard me make that call? How would I explain it?

'Mum said you spoke last night,' he says with his back still to me. 'She was saying that Cass is getting pissed off about the lack of details from us about the wedding. Can you send her a message?' He turns around and I glare at him.

'She's *your* aunt,' I say, unable to hide my annoyance that *this* is the sort of thing everyone thinks I should care about.

He frowns. 'Yes, I'm aware of that. But I'm a bit busier than you right now...' He takes off his shirt and puts on the fresh one, shoving the old one into the wash basket.

'What does she even want to know?' I ask.

Nath shrugs as he puts on his trousers, making him nearly tip over, but this comedy moment is not enough to shake me out of this mood.

'Dunno, give her a call. You know what she's like; she's probably got ten different outfits to choose from and wants to pick the one that will upstage Mum the most.' He flashes me a smile, as a peace offering.

Amanda and Cass's rivalry is something we always laugh about; they've got worse the older they've got. Cass is a typical stay-at-home wife and never failed to use this to look down on Amanda. Despite how close they are, the sibling rivalry never stops. I often moan about being an only child but seeing them, this late in life and still engaging in these childish battles, makes me wonder if I'm the lucky one.

'I honestly couldn't care less what either of them wear,' I say, sitting up on the bed as Nath goes to walk out. He stops, his hand on the door frame, back to me, then turns around slowly.

'Why do you have to say it like that?'

'Like what?'

'Like everything about the wedding is so utterly stupid to you. Like it's beneath you.' Hurt blooms across his face and I want to take it back but instead dig my heels in.

'Because most of it is!' I throw my hands up. 'Grown women getting het up about what dress to wear when there's real stuff going on—'

'Like what?'

'What do you mean "like what"?' I snap. 'Like I've got nothing else to think about except this wedding, is that what you mean?'

'Well,' he says. 'You haven't!'

I scoff. 'Fuck you, Nathan,' I say and storm out of the room, pushing past him down the stairs.

I hear him follow me and know I should stop – I should turn around and tell him I'm sorry, that I'm being unreasonable and I do care about the wedding. How can he think that I've got nothing else going on in my life except being a blushing bride?

'You're not working, you've got nothing to do and yet you're still showing absolutely no interest in the wedding. Did you even call the photographer yesterday?' he snaps from behind me and I close my eyes. Fucking photographer.

'I'll call her later—'

'But why didn't you do it yesterday? What kept you so busy that you couldn't make one simple phone call?' He pulls me around to face him and his eyes search my face. I open my mouth, but then close it and shrug.

'I'm going back to work,' he says.

'Nath, wait,' I say, reaching for him. He stops, looks at me, desperately hoping for an apology.

'I'm sorry,' I say.

He instantly softens, pulling me into him. 'I just want to understand what's going on, Sal. I feel like you're not telling me something.' He holds me by the shoulders and looks into my eyes. 'Do you... Do you not *want* to marry me?'

'It's not that, of course it's not that,' I tell him, hating myself for making this man – this good man who I've loved my whole life – doubt himself like that. 'Something happened yesterday...'

'What?'

I nod towards the living room to give myself a second to think; he follows me through, and we sit on the sofa next to each other. I can't tell him about Sally, but I need to tell him *something*.

'I met this girl on the beach... Tiffany,' I say, and he frowns but gestures at me to go on. 'She was only a kid, fifteen or so. Her friend's flat had been broken into.'

'Right...' he says, waiting for the follow up.

'She wouldn't call the police.'

'Why not?'

I shrug. 'She said she wanted to try and figure out what was going on herself first.'

'Probably means she was involved,' he says, so sure of himself like there's no other reason someone wouldn't want the police at their door. 'What did her mate say?'

'The woman in the flat wasn't there. Tiffany, well... she doesn't know where her friend is.'

Nath screws up his face and stares at me like I'm insane. 'As in, the woman whose flat has been wrecked is also missing?'

I nod.

'And you just met this girl on the beach and she told you all of this?'

I sigh and shrug. 'I must have a trustworthy face.'

Nath frowns. 'I can talk to my mate down there, find out about it if you're worried.'

'No,' I say, too quickly. 'No it's fine. She said she'd report her missing—'

'Sal, if you know a kid is missing—'

'She's not a kid, the friend – she's an adult. The girl said she was about my age. They live in the same tower block.'

'Right,' Nath says. 'So, this girl's friend's flat has been broken into and the friend is now missing.' He says, establishing the facts, and I nod but don't look at him. 'And no one has told the police.'

'Yes,' I say. 'Well, as far as I know...'

'Fucking hell, Sal. How do you get involved in these situations? I thought you were spending the day clearing your head.'

'It's not my fault!' I cry, sounding like a teenager myself. 'I didn't mean to get involved.'

'You have to tell the police,' he says, standing up and dusting off his trousers. 'I've got to get back. But call Avon & Somerset. Just give the girl's name and let them follow it up. Don't get involved more than that. Okay?'

I tell him I won't, feeling like I've had a telling off from my parent.

'Seriously, Sal,' he says as he reaches the door. 'The last thing you need right now is more drama. We're about to get married, can you please, please just focus on that?'

He leans in for a kiss and I smell the woody scent of his aftershave on his freshly shaved neck. I tell him I will, I promise, I will.

CHAPTER TEN

I know I promised Nath no more drama, but I have to go and meet Alex at Sally's flat. I need to confront him and find out what he knows about her disappearance. Then, if I can't find anything that will help her, I'll ring the local force down there.

I've had no reply to my email telling Sally I know about Alex and I'm trying my hardest to not be annoyed by this. Why can't she see I'm trying to help her? It's like our friendship has meant nothing to her at all, but perhaps that's because it was built on lies.

I've not heard back from Tiff either, after I messaged to say I'd be back in Weston today and asked if she was around. Seeing that she's read the message but not replied, I feel a little silly. She's a teenager, not my friend and why am I even trying to get her involved in what could be something dark? But it's too late now.

Nath's not going to be home until later this evening so I've got enough time to get down to Weston, confront Alex and get back home before Nath even knows I've been gone. Plus, I've also already rung the photographer for the wedding – she wasn't super happy I'd ignored her for days and now wanted an immediate answer from her, but at least that's one thing ticked off my list that will keep Nath happy. I've also texted his aunt and said she can call me later tonight about her outfit if she wants to.

After a strong cup of coffee, I'm pepped up enough to get through the day. I'll figure this thing out with Sally and get home without anyone knowing I was gone. No issues. As I grab my sunglasses from the bookshelf by the front door, I feel empowered. I'm making things happen; I'm finally on top of it.

—

Two hours later, I pull up outside Sally's tower block and push my sunglasses on top of my head to keep the hair from my face. There's no need to familiarise myself with the area any more or make up a fake story for why I'm in Weston. If I run into Sally, I'll be so bloody relieved that I'll happily tell her I've effectively become her stalker in the last few days. It's gone past the point of polite embarrassment now and I just need to know where – and who – she really is.

I check the time: twelve thirty. Fifteen minutes until sleazy Alex the letting agent arrives. There's no sign of Tiff hanging around like yesterday; perhaps she's gone back to school. That's for the best, even if it means I won't have a sidekick for the day.

I make my way up the dark staircase, breathing through my mouth to try and avoid the smell of stale urine as much as I can. When I reach the fifth floor, I march confidently down the corridor to Sally's door. I push it open, noticing the lock is still very much broken and am greeted by the same sad wreckage of a home I saw yesterday. I don't know why I expected any different but as I survey my surroundings, I realise part of me was hoping to find Sally here, sitting on the sofa, safe and well with Grace happily playing with her toys on the rug beside her. Stupid. The life Sally painted for me was clearly always a fantasy.

But still, this isn't the time to berate myself for falling for her lies – Alex will be here in five minutes. Taking out my phone, I snap some photos of the living room in case it's useful for the police later. I *will* phone them today. Unless Alex has some information that means I don't have to. My stomach rolls when I think of what I'm about to do. What if he turns nasty? My earlier positivity is seeping away the longer I spend in this depressing tower block.

In Sally's bedroom, I pull the mattress over so that it's at least facing the right way and sit tentatively on the edge, my feet curled up beside me. The room smells of damp and something else I can't place; is it urine? I sniff around the room, imagining whoever broke in taking one giant piss all over Sally's belongings and my nose curls. But then I hear something from the chest of drawers that might explain the smell and my stomach twists.

'Oh no,' I say, opening the drawer. 'Oh god, look at you.' The small, pained cries hurt my heart as I look into the huge blue eyes staring up at me like I'm their saviour. 'Let's get you out of there.'

I hold the cat close to me, its small skinny body warm and wet with its own urine.

'Poor little thing,' I say as the cat continues to mew at me. 'How did you get in there?' It burrows its head into the nook of my arm and starts vibrating with happy purrs. We don't deserve animals. We really don't.

'Sally?' I hear a man's voice shout from the front room, quickly followed by an exasperated muttering of, 'What the fuck?' as he takes in the wreckage.

I put the cat down and it scuttles under a pile of clothing in the corner of the room. Good, I think, you hide.

'In here,' I call back trying to mimic the voice of a woman I've never actually heard.

As Alex's footsteps approach, heavy and uneven, my heart hammers under my ribs. What the fuck am I doing? This is quite possibly the most stupid decision I've ever made but as the door pushes open I pull my shoulders up to stand at my full five-foot-five-inches and glare at the man who stands before me.

His ratty eyes dart over me, a startled and indignant sneer that's quickly covered by a slick smile.

'Oh,' he says, clearing his throat. 'I was expecting Sally. Who are you?'

'I'm Sally Jones,' I say with a smile. 'You're Alex Layton. Sally's landlord.'

He narrows his eyes and we stand a metre or so away from each other saying nothing.

When the silence gets too much, he says, 'You're not Sally.'

I shrug. 'And you're not a very good landlord, but there we go.'

'What the fuck is going on?' he says, his arms thrown up like a petulant child. 'Where's Sally? Who wrecked the flat? If it was you then you're going to need to pay—'

'And how would you like that payment, Alex?' I say, my voice strong and stable now. I'm in my element as I watch this embarrassing excuse for a man rock back and forth on the heels of his cheap shiny shoes. His horrid little mouth opens and shuts like a fish, the wispy moustache he's trying to grow wavering in the gust of his hot breath.

'How long have you been demanding sex in exchange for rent, Alex?' I ask, calmly.

'This is ridiculous,' he says, turning to leave. 'I'm not listening to this shit. I'm calling the police.'

'Good luck,' I say. 'Seeing as I am the police.'

That does it; he stops, doesn't turn but stands stock still with his back to me and I enjoy the power I have now – the ability to stop this pathetic man in his tracks.

'Look,' he says, turning slowly. 'I think we've got off on the wrong foot…'

I smile and shake my head. 'I don't think so. You've been abusing your position to extort sex from another person. That's illegal, Alex.'

His face drains of colour but then he narrows his eyes and studies me. His glance skitters over my body in a way that makes me want to shower immediately, and then a small smile curls his thin lips. 'You're not a police officer. Show me your badge.'

I tilt my head and shudder, as if that will shake his gaze from my body. 'Alright,' I say. 'I'll admit, I'm not here on official business. But I am police, and more importantly, I'm a friend of Sally's.'

'Fuck off,' he says, his earlier fear long gone. 'Dumb fucking bitch. You tell Sally if she doesn't pay up for the rent and this fucking damage I'll—'

'You'll what?' I say, pulling my phone out of my pocket to show I've been recording this whole conversation. He stutters and I flash a smile. 'Come on, Al. Don't be shy. What will you do?'

'Fucking hell,' he mutters under his breath, defeated. 'Alright. Look, what do you want? Turn that off and we can talk.'

I shrug, switching off the recorder like it's no big deal to me either way. As much as I've enjoyed making Alex squirm, what I need is information and there's no way he'll talk on the record.

I gesture for him to go into the living room and then follow behind. After clearing the sofa of the various bits of crap that have been thrown on it, he sits down as I shove my backpack against the front door to keep it shut while we talk. It has the exact unnerving effect on Alex I was hoping it would, and he shrinks back on the sofa.

'Sally and I have always had an agreement. It works both ways, alright? If it wasn't for me she'd be out there,' he points to the hallway, though I assume he means homeless.

'What a good guy you are,' I say with a head tilt. 'Do you know what happened here?' I ask, gesturing to the destruction around us.

He shakes his head. 'Do you?' he retorts and I shake mine.

'When was the last time you saw Sally?'

He shrugs. 'Couple of weeks ago. Whenever the last rent was due...' He flashes me a wolfish smile and I struggle to contain my disgust.

'Have you got a photo of her?'

He frowns. 'Why would you need a photo of her? Thought you were her mate.'

'I am,' I say, my cheeks reddening. 'Look, I do know her. But we've only spoken online.'

He laughs, like I'm the one who should be embarrassed in this situation.

'Alright, *mate*,' I say. 'Please remember it would literally take one phone call to your workplace to ruin your whole life.'

The laughter stops and he gets out his phone.

'You can't really see her face,' he says without shame, holding up the screen to me.

It's a shot of a woman in her bra and pants – not matching – not the kind of shot you see on TV shows where the woman grins with delight at her own beauty. He's right, Sally's face is barely in the photo. It's obviously been taken in a moment of desperation, not mutual excitement. She's thin, the lines of her ribs visible, but her belly is soft and saggy. This is not a body forged in the gym; this is a body deprived of nutrients and love. She's young too. It's hard to say exactly how young but likely in her twenties. The whole image makes my stomach turn and as I look back up at Alex it's clear even he knows how seedy having this photo is. He pockets the phone and looks down to hide the shame on his cheeks.

'Probably best if you delete that,' I say, and Alex rolls his eyes but does as I ask then holds his phone up to confirm it's done.

'Gone,' he says, the cocky confidence slinking away. 'Look, I really don't know what's gone on here. I care about Sal,' he says, and I shake my head in disgust. 'I do! I know it's not the most conventional relationship—'

'It's not a relationship. You're literally forcing her into sex.'

He waves his hand like this is neither here nor there. 'She's got issues, alright? She's the type of girl who you just end up doing this type of stuff with. I mean, look!' he says, looking around the wrecked living room as if this is proof of Sally's 'type'.

'How long has she lived here?' I ask.

He shrugs. 'Years. Dunno, at least four. She used to pay on time, but the last six months she… I don't know. She never had any cash. My dad wanted to chuck her out but I said I'd sort it.'

He says this like he's her knight in shining armour and I use every ounce of energy I have not to scoff.

'What about Grace?' I ask and he looks at me blankly.

'No idea who that is, love.'

'Her daughter?' I say to more blank stares. 'Grace. Her daughter... She's five?'

He shakes his head. 'If she has a daughter, she doesn't live here.'

I bite my lip, this information panicking me more than it should. It was already obvious Sally's life wasn't as she'd portrayed but the revelation that Grace doesn't exist is so hard to take. Why would she lie about that?

'What else can you tell me about Sally? Where does she work? Do you know any of her friends?'

Alex throws me a look of derision, almost like he feels sorry for me, but then he stands up and brushes down his cheap suit like being in Sally's flat might have left some sort of debris on the synthetic material.

'Look, I barely know the girl. She's just a tenant. She used to work in the pub down the road, no idea if she still does and I don't know any of her mates. Are we done? Because I've got things to do.'

'We're done,' I say, standing up to face him. 'But I'll be watching you.'

He scoffs.

'Alex, I'm deadly serious. I deal with men like you every day and trust me, your kind think they're all that until they end up in prison and realise suddenly, *they're* the little bitch everyone else uses for their own satisfaction.' I stare him dead in the eyes and wait for him to break.

He says nothing, then nods and walks out of the flat.

I flop back on the sofa, adrenaline coursing through me, making my whole body shake.

CHAPTER ELEVEN

Leaving Sally's flat, I'm dejected. Have I learned anything more about Sally today or have I risked putting myself in a vulnerable position with a man I know to be a predator for nothing? The photo Alex showed me hasn't exactly helped; I still wouldn't recognise Sally in the street if she walked straight past me. Not unless she was wearing her bra and pants, anyway.

At least he confirmed Grace doesn't live with Sally – or perhaps doesn't exist at all. Something about this discovery makes me uneasy. What sort of a woman lies about having a child? Clearly, Sally isn't the person I thought she was. But I am softening towards her. The lifestyle she leads is one of desperation. Perhaps her lies to me were her only escape from it all. Is that why she lied about Grace, too?

I pull the flat door closed as much as I can and walk back down the corridor before remembering the cat. I can hardly leave it in that flat alone, can I? I turn on my heel and go back in.

'Here puss, puss, puss...' I call, making a *tsk* noise with my tongue. 'Where are you?'

I scan the room; a pile of clothes twitches in the corner. As I move a dirty jumper to the side with my foot, the cat's little face appears. It stares up at me then slinks out from under the clothes, weaving in between my legs as it purrs against me.

'What am I going to do with you?' I say to the cat.

I get on my phone and google 'found lost cat' and read an article from Cats Protection. It says to take them to a vet to scan for a microchip. Thing is, the cat isn't lost, is it? It's home, presumably. Sally has left it behind. This strikes me as incredibly callous. Again, I have to ask myself what sort of woman she is. How could you leave your cat behind to starve to death? Not that the cat looks in any danger of that, with a big belly hanging down between its legs. But still. She's left it here all alone. That takes a certain type of person.

Or, it means she had no choice but to leave. I think back to her last email warning me off. Again, perhaps I'm judging her too harshly.

Either way, I can't leave the cat here alone. I search around the messed-up room for a cat carrier but there's nothing, then eye up my backpack. Could I? I look at the cat, who looks back at me as if to say, 'you're the grown up here, not me' and then unzip the bag. The cat immediately sticks its head in and purrs before climbing inside. I gently zip it up as I watch the cat's face disappear. The animal is docile as I pick up the bag, its purr vibrating against me. Worried it'll suffocate in there, I leave a gap open and the cat pops its head through, happily purring away as we walk out of the flat.

—

As I reach the front door of the tower block, I hear a familiar voice.

'What the fuck are you doing?'

I whip around. 'Tiffany!' I break into a smile but then it slides off my face when I notice her expression. 'What's wrong?'

She scoffs and shakes her head at me like I'm some sort of delinquent, then lurches towards me and grabs my bag.

'Oi!' I shout but she snatches it out of my reach then wrenches the zip open.

'You're fucking mental,' she says as the cat jumps out and sits beside her feet, licking its paws nonchalantly.

'It's Sally's cat, I couldn't leave it alone—'

'It's not Sally's cat, you mad bitch. It's the block's.'

'What?'

'He lives here,' Tiff says, as if this is the most obvious thing in the world and I am a total idiot.

Of course, I do feel like a total idiot now that I think about it. 'He was in the flat…'

'He goes in all our flats,' Tiff says her eyes still narrowed and furious at me.

'I didn't know—'

'Nah, well you wouldn't. Would you?'

I tilt my head, unsure of what's happening. 'Have I done something to upset you, Tiffany?'

'What, other than trying to steal our fucking cat?' she spits.

'Yes, other than that…'

The cat finishes licking itself and strolls off down the corridor, happy as anything.

'You're not right. You've got one of them saviour complexes, ain't you? Probably recording all of this for TikTok so you play out the video over some shit music and show what a good person you are.' Tiff curls her lip at me and shakes her head. 'Coming in here, like you give a shit, then taking things that don't belong to you.'

'Well, I don't post anything on my TikTok, for a start,' I say trying to lighten the mood but her scowl tells me this is a mistake. 'Tiff, I told you it was a genuine mistake. I thought the cat would starve to death without Sally here to look after it,' I add more sincerely.

'Not "*it*",' she hisses, 'him. His name's George,' she says and despite myself, a laugh bubbles in my throat. 'The fuck are you laughing at?'

'Oh come on,' I say. 'This is silly. I wasn't exactly *kidnapping* George; I was just trying to make sure he was safe.'

'That what you told yourself when you stole Sally's kid too?' Tiff spits at me, fury burning on her cheeks.

'What?'

'You know what,' she says, crossing her arms angrily.

'I really don't, Tiff. I've never even met Grace.'

She narrows her eyes at me. 'I know what you are,' she says, taking a step forward, which makes me instinctively take a step back.

'What am I?'

She sniffs, then leans so close to my face I smell her warm breath on my cheek. 'A pig,' she whispers into my ear, then for good measure adds a pig-like snort. I take another step back and try to regain some composure.

'Yeah okay, I work for the police. I was going to tell you that last time I was here but...' I stop, not knowing what to say. 'How did you find out?' I ask.

Tiff shrugs. 'Googled you, didn't I? Can't trust some mad bitch you meet on the street, can you?'

I tut at her words. 'Look, Tiff. I'm sorry. About George, and about not telling you I was in the police. But I'm not here on police business, I promise. What I told you about Sally was true, I really am just a friend who's worried about her.'

'Maybe she found out you were a pig and didn't want anything to do with you after what you lot did to her.'

I frown. 'What did we do to her?' I ask.

'You stole her kid! Her Grace. No wonder she pretended she still had her. Imagine what it feels like for someone to steal your baby away.'

'Tiff, I promise, I really have no idea what you're talking about.'

—

Begrudgingly, Tiff has let me into her flat so we can talk about what's going on. It's the exact same layout as Sally's but that's where the similarities end. The place may not be neat, but it's lived in and cosy. On the walls of the living room are huge canvases with family photos; Tiff still looks like a moody teen in them but next to her brothers and sisters, there's an innocence about her. Her mum is young, probably my age, which feels a bit weird but there's no reason why I couldn't have a child of Tiff's age myself, so I turn away from the photos and back to the job at hand.

'So, will you please tell me what you mean about Grace?' I say, as Tiff sits on the worn leather sofa beside me.

She scowls and searches my face as if debating with herself whether she can trust me. 'After you left, I did some research,' she says and I nod, encouraging her to go on. 'I found a mate of Sally's, Layla. She lives on the same floor.'

'Layla,' I say, committing the name to memory. 'How did you find her?'

Tiff shrugs like that's not important. 'She doesn't know where Sally is either, but even if she did, I wouldn't tell *you*.'

'Right,' I say reminding myself not to push Tiff too hard or she'll clam up altogether. 'Why not?'

' 'Cause you lot can't be trusted. Layla said she knew Sally was missing but wouldn't ring the pigs because Sal hated them after what they did to her.'

I hold my hands up, urging her to go on but she just shrugs like that's it.

'Tiff, did she tell you exactly what she meant? Did Sally have her child taken off her by social services or something?'

'Nah,' Tiff says, shaking her head. 'Not official, like. Layla said it was all corrupt.' She nods knowingly and I wonder how much of this is true and how much is Tiff turning an already dramatic situation into something far worse.

But what if it is true? I don't believe for a second the police 'stole' Sally's child but she may well have had her little girl taken off her for a genuine reason. Was it my mention of both Nath and I working for the police that made Sally cut contact with me? But no, that doesn't make sense because the newspaper article she had about me was years old. She must have always known I was in the police, even if I didn't tell her directly.

'I don't understand,' I say, as if Tiff will have any answers for me. 'Can you take me to Layla, please?'

Tiff shakes her head. 'No chance,' she says. 'She don't want anything to do with pigs either.'

'Tiff, can you please stop calling me a "pig"?' I rub my face with my hands, wondering where the hell we go from here. When I look back up, Tiff is staring at me but I can tell some of her anger has abated so I try again.

'Is there anything else Layla told you that might help me work out what's going on? Was Sally in some sort of trouble?'

She shrugs. 'Layla said not. I mean, no more than normal.'

'It would be much easier if I could talk to her directly—'

'What are you even doing here?' Tiff asks.

I wonder if Layla knew about Alex and whether this constituted the 'normal' amount of trouble for Sally to be in, but I know Tiff is not the right person to be discussing this with. Despite all her bravado, she is just a kid. So, I lie.

'I thought I had a lead on where Sally might be but it turned out to be nothing useful. I was hoping I'd be able to find her, without having to tell the police about the break in.'

'You didn't tell the pigs?' Tiff says with her eyes wide.

I shake my head.

'Won't you, like, get in trouble for that?' She twists her hair around her fingers, a move that shows how young she really is.

I shrug. 'Yes, yes I probably will. But still, a bit late to be worrying about all that, isn't it?'

Tiff laughs and I smile, then tense.

Because the familiar sound of sirens is ringing in the background. Tiff and I lock eyes.

'You're full of shit,' she spits as she pulls me up from the sofa and forces me out of the door.

'I didn't call them!' I shout as she slams the door in my face but it's too late. She's hidden safe behind the door and as the sirens get louder, I realise that I should be hiding too.

CHAPTER TWELVE

As I step back out onto the street, a police car zooms past me in a blur of blue lights while the ear-aching sirens scream out.

I let out a sigh of relief. They aren't here for me or for Sally. But still, it's a stark reminder of what is at stake here if I don't sort myself out. How would it look to be caught in Sally's flat? What would my inspector say if she found out I knew a woman was missing and instead of doing the right thing, the thing my job requires me to do, I'd played amateur detective? This is crazy. It has gone too far. I look at my phone in my hands, with 111 on the keypad ready to be dialled. There's nothing else I can do. I must tell the police about this. Perhaps if I explain why I didn't call it in earlier they'll understand.

No. They won't.

My inspector already thinks I'm unstable enough to force me to go on leave. This will not help my case and I can't bear the idea of being forced out for any longer. I won't ever be able to help anyone again. And what then?

I will do the right thing. I will call it in. But not to the police.

I walk to the corner of the estate where Sally lives, sure there was a phone box here the last time I drove past. There it is! Oh great. It's been turned into a defibrillator box. Looks like I'll have to call from my mobile after all.

I delete the police number and key in another I know off by heart — perk of the job — standing inside the old phone box to give me some privacy.

'Crimestoppers. How can we assist you today?' The warm voice of a woman floats down the phone and I imagine her sitting in a bunker like mine at work, a cooling coffee by her side, her colleagues rolling their eyes at her from across the room as they listen to the general public's woes.

'Hi, I'd like to erm...' I fumble my words, unsure of myself on the other end of this conversation for once. 'Report a crime?'

'Are you happy to give us your name or—'

'No, I'd like to stay anonymous, thank you.'

'No problem. Is this an existing crime with a crime reference number, do you know?'

'No. It's a friend... of a friend...' I say, wishing I'd rehearsed this before making the call. 'Her flat's been broken into and she's missing.'

'What's her name?'

'Sally Jones,' I say.

'And the address of her flat?'

I reel it off and wonder briefly if I should be making the call from further away, as if the police are going to swoop in any minute and bust me. But I know that's ridiculous; they don't have the resources to take action like that, not for a missing person case.

'And when did you last see Sally?' the woman asks.

'I... I haven't seen her. But I know she was seen by another friend a week or so ago.'

'Do you have this friend's name?' she asks.

'No,' I say. I can't tell them about Layla, because that will lead them to Tiff, which will lead them to me. 'Look, that's all I can say. Something is wrong – Sally is in trouble. Can you just look into it?'

The woman starts to reply, tells me the more information I can give the more likely they are to be able to help my friend but I know what I've given them is enough to make a start and so I hang up.

–

I drive back up the M5 on autopilot. Will Crimestoppers have passed it onto the local police yet? Have they been to Sally's flat, seen the wreckage and started looking for her? Maybe they've spoken to the neighbours. Will Layla tell them about me – this stranger that's been sniffing around? A sweat forms on my upper lip. This is going to lead back to me. All I've done by involving Crimestoppers rather than the police directly is delay the inevitable. I should come clean now before it goes any further. That's what Dad told me he'd always advise his suspects. To 'fess up as soon as you can. The more you try and hide, the worse the consequences. But my reputation is already in question with the higher-ups at work right now; can I risk making things even worse?

I wish Dad were here to talk this through. He'd be shocked at my stupidity, but he'd sort it out. I know he would. Dad might not be here any more, but there's still one person I know who will always help me out of a mess.

By the time I pull into the familiar driveway, the weather has turned gloomy and the sunglasses I've worn for the whole drive are making everything appear like night-time despite it only being just past five p.m. Nath has texted a few times and said he'll be home for dinner tonight and suggested we get a takeaway. I'm nervous to see him but ignore the feeling as I get out of the car and walk up the driveway to knock on the reassuringly weathered red front door.

'Sally?' Amanda appears at the door, her face creased with concern. 'What on earth is wrong?'

I burst into tears the moment she pulls me in for a hug.

'Come in,' she says, pulling me over the doorstep. 'Whatever it is, we can sort it out,' she promises, and this is exactly why I've come here. No one can fix things like Amanda.

—

Amanda has made us a pot of strong coffee and placed it in the middle of her oak kitchen table. The smell of it is bringing me back to life and I've wiped away my tears with her balsam Kleenex tissues she always keeps next to the fruit basket. Today, the house is spotless — a sharp contrast to the cosy, family chaos it exuded back when we were kids. I loved spending time here amongst it: our house was too quiet with just Dad and me. We always felt more like a family sat around Amanda's kitchen table than our own. I look across at the empty seat that used to be Dad's, feeling a twang of pain at his absence.

'So, do you want to tell me what's going on?' Amanda says, her dark eyebrows furrowed in concern. 'Is it about the wedding?'

I lift the warm mug to my lips and look down.

'Nath mentioned you'd been a bit uninterested in it all lately...'

'Nath said that about the wedding?'

Amanda nods and hurt sprawls across my chest.

'You know what he's like,' she says. 'All tough on the outside — happy to play the big man at work — but he's a sensitive soul deep down. You're not having second thoughts, are you?'

'No, of course not. I love Nath,' I say then let out the most enormous sigh. 'I've got into such a mess.'

Amanda doesn't say anything, and we sit there in uncomfortable silence. I wonder if this is how she was in suspect interviews back in the day, giving them enough silence to fill with their confessions. Dad always said she was the best interrogator around – even better than him.

'I've got caught up in something,' I say when I can't take the silence any more. 'Something bad.'

'I'm sure whatever it is, it can't be that bad.'

I make a murmuring noise in my throat and shake my head.

'Look, Sally, whatever it is, I'll help,' Amanda says, reaching across the table to grip my hands. 'You know I think of you like a daughter, but I can't help you if you won't tell me anything. Just tell me. Is there somebody else?'

I pull my hands back in alarm. 'Someone *else*?' I say. 'Of course not! I'd never do that to Nathan.'

Amanda shrugs, as if an affair is neither here nor there. 'What is it then?'

I twist my hair around my fingers anxiously, not knowing where to start.

'You remember that friend I told you about the other day at lunch…'

'Sally,' Amanda says with a raised eyebrow again, as if she still believes this 'friend' who shares my name might be entirely fictional.

'Yes,' I say, 'her.'

'What about her?'

'Well, I went to see her.'

Amanda narrows her eyes, as if this isn't at all surprising. 'In Weston-super-Mare,' she says.

'How did you know that?'

Amanda sighs and tilts her head. 'You told me you'd got a rogue email from a carpet company in Weston and that's how you met this "Sally", remember? Then you disappeared off down there the next day. Hardly takes a genius to work it out, love.'

'So you knew why I'd been down there the other day?' I say, thinking back to our phone conversation where I lied.

'I was a detective for thirty years, Sal. Of course I knew.'

I scratch at the back of my neck, wrong-footed. If she knows, does that mean Nath does too?

As if reading my mind, Amanda says, 'I didn't say anything to Nathan. I thought you'd tell him yourself when you were ready. So, what's going on with this friend that's got you so worked up?'

She reaches forward and pours us both more coffee though it's not needed in my case; caffeine is already buzzing through my tired veins, making me shake.

'I hadn't heard from her for a while then I got this email saying she hadn't paid her rent and was going to be evicted,' I say, watching Amanda's face for clues of how shocking she finds this, but she gives me nothing, just nods at me to go on. 'So I decide to drive down there but when I do, the flat, well…' I stop, knowing that from here on I could get into trouble.

'Well, what?' Amanda says, her ice-blue eyes wide and waiting for me to continue.

'It had been broken into. Trashed, really. It didn't look like a burglary.'

'How so?'

I shrug. 'I didn't get the impression anything had been taken.'

'But you'd not seen the flat before, so how would you know?' she asks in that logical matter-of-fact way she has of dealing with everything.

'Well, Sally doesn't seem like the type of person who had a lot worth stealing.'

'I see,' Amanda says, picking up her coffee again and taking a sip before looking back to me thoughtfully. 'Go on.'

I take another deep breath. 'I mean that's it, really. She's missing, something has happened to her and I've no idea what, but it must be something bad.'

'You've not reported it,' Amanda says, a statement, not a question.

'Not exactly...'

She raises her eyebrows at me, and I find myself talking again.

'I rang Crimestoppers,' I say and she tightens her lips, a slight shake of her head telling me I may as well have not bothered reporting it at all. She's never held a particularly high opinion of the organisation.

'So, what were you doing down there again today?' she asks.

'I wasn't... I, erm...' I bungle my words like a thief caught out. 'How did you know I'd been there again today?'

'Well,' Amanda says, standing up and walking around the kitchen with her coffee still in hand. 'You phoned me from the car asking if I was in. It sounded like you were on a motorway and you turned up an hour or so later. So logically, I would make a decent guess that's where you came from. Am I correct?'

I say yes, looking down at my hands on the table like a criminal in an interrogation room. I don't know why I tried to hide anything from Amanda, she can obviously see right through me. She always has.

'So, aside from being worried about this missing "friend",' she says, 'your problem is this: one, you've seen evidence of a potentially violent crime and haven't reported it, which directly violates the force's code of ethics. Two, you've gone to Crimestoppers which could, I imagine, eventually lead them back to your involvement. So the force will, in all likelihood, find out. And three, you haven't told Nathan a thing about any of it.'

I look up and nod. Hearing it laid out like that sounds even worse than I'd been imagining. As tears threaten, Amanda sits back down and pulls her chair closer to me.

'God Sal, you don't half get yourself into some messes, don't you?' She rubs my arm like she used to when I'd fallen off my bike, and that's what finally makes the tears fall.

'I'm sorry,' I say. 'I don't know what I was thinking.'

'You were just trying to do the right thing. Help out a mate. There are plenty worse things you could do, trust me,' she says in a soothing voice as she continues to rub my arm. 'Don't worry, we'll sort this out.'

'There's something else,' I say, and Amanda raises her eyebrows as if to say, 'really?' and I continue before I lose my nerve. 'Another reason I didn't report it. Sally… She doesn't want to be found.'

'What do you mean?' Amanda asks.

'She sent me an email,' I admit. 'Warning me off. She said if I told the police, things would get worse.'

Amanda frowns, then shakes her head. 'Nothing is ever made worse by doing the right thing, Sal. This is a police matter, you know that.'

—

After Amanda has reassured me that there's no way I can keep pretending this is something I can handle on my own, I've been left in the living room. Amanda has gone upstairs to her office; despite not working any more she still has one room dedicated to her old working life, where a large black monitor sits in the corner of the office alongside a dusty printer and sturdy office chair. Nath makes fun of her for it, asking what she can possibly need a set-up like that for any more, but I think part of it is to remind herself she was once a valued professional.

I can hear her walking back and forth across the carpeted floorboards upstairs as she makes calls. I've no idea what she's doing but after I told her everything I knew about Sally she said she'd see what she could find out. Deep down, I knew this would happen from the moment I turned up on her doorstep. I've always been able to go to her to solve my problems, no matter how bad.

In my pocket, my phone buzzes and Nath's name is on the screen calling me.

'Hi,' I say.

'Hi, babe,' he says, his voice crackling as I imagine him driving down the A429 on the way home from a shift. 'I'm on my way back.'

'I'm at your mum's,' I say.

'Oh,' he says. 'What are you doing there?'

I hesitate, then say as casually as I can, 'Just thought I'd pop in. Got a bit bored on my own. How long will you be? I can leave soon.'

The sound of a blaring horn cuts through his words and I hear him snapping 'fucking idiot', presumably at another motorist. 'Sorry, babe. Honestly, I know you hate it when I say it, but I really do think anyone over the age of seventy should have to retake their driving test. They're fucking menaces.'

I laugh because I know he wants me to. He's in a good mood and I'm not going to be the one to ruin it.

'I'll be forty minutes or so,' he says. 'Traffic's completely backed up at Longbridge. But don't rush if you're happy at Mum's. I can order us something for when you're back. What are you in the mood for?'

The idea of food turns my stomach but I tell him Chinese because I know he's always happy with that and it'll get him off the phone quickly.

'The usual?' he asks and I tell him yes, even though imagining the gloopy brown-and-orange-hued curry sauce dripping over a pile of egg-fried rice and chips makes me want to gag right now. 'See you soon then, love you,' he says.

'Love you too,' I say, before ringing off.

'That Nathan?' Amanda asks from the doorway; I hadn't noticed her come in but I nod. 'Did you tell him?'

'No,' I say. 'I told him I wanted some company so popped by here.'

'Good,' Amanda says, coming to sit on the sofa opposite me. 'Probably best we leave Nath out of this for now.'

I agree, relieved but curious. 'Why?'

Amanda smiles. 'Like I said earlier, Nath's a sensitive soul and I don't know if he'd be comfortable with me bending the rules for you like I've had to do...'

'What do you mean?'

Amanda brushes down the sofa cushions beside her, although there's not a speck of dust anywhere in the house. 'I've had to call in a few favours to get information on your friend "Sally".'

'Oh god, Amanda, I'm so sorry. You shouldn't have to do this for me...'

She waves away my concerns. 'Like I said, I was owed some favours. It's always useful to have friends in high places owing you things.' She winks and I see a side to her I've not been allowed to before. Turns out she might not be quite so rule-abiding as I thought; perhaps that's a façade she's successfully shown to the world for all these years.

'Thank you,' I say. 'I really do appreciate it.'

'Don't thank me yet,' she cuts in. 'You don't know what I've found out.'

The sick feeling hits me in the pit of my stomach again, so I lean forward to ignore the swirling.

'I'm afraid it's not good news,' Amanda says, and I tense. Have I acted too late? Has Sally already come to harm from whoever she's running from? Perhaps my anonymous call earlier triggered an investigation and right now Sally's cold dead body is lying in some morgue all alone.

'What is it?' I ask, unable to wait another second to hear the news I've been dreading.

'Well,' Amanda says, clearing her throat and sitting back against the sofa. 'I asked an old friend who works in intel down there to look up Sally's name, see if she had any sort of record to start with but there was nothing,' Amanda says, and I let out a breath.

'Well, that's good,' I say hopefully.

She shakes her head. 'No, Sally. There was nothing, nothing at all.'

I frown. 'But that just means she hasn't come into contact with the police as a victim or suspect; that isn't that unusual,' I say, but don't add what we're both thinking, that for someone growing up in a block like Sally's, not having any police involvement is incredibly unusual.

'It goes further than that,' Amanda says. 'This friend did extra checks and the name "Sally Jones" doesn't bring anyone up at all matching the woman you've described.'

'Oh,' I say.

'Your "Sally Jones",' Amanda says, using air quotes on the name, 'doesn't exist.'

CHAPTER THIRTEEN

'I don't understand,' I say to Amanda after a moment of silence trying to absorb this information. 'She does exist. We've been emailing for years. She's got a flat in her name, she has appointments that I've seen… she's a real person.' I sound desperate, and Amanda looks at me with pity.

'I know, sweetheart. I'm not saying your friend isn't real,' Amanda says patiently. 'But it looks like she's been using a fake name.'

I shake my head. 'Why would she do that?'

Amanda shrugs. 'Your guess is as good as mine.'

'But why *my* name?' I say, to myself as much as Amanda. 'Why choose that?'

'It will just be a coincidence,' Amanda says. 'Don't take this personally.'

'But it is personal,' I say. 'She took my name then befriended me…'

Amanda shakes her head. 'No, Sal. She took *a* name, probably any name – maybe she always liked the name Sally and, well, Jones is a common surname. This doesn't have anything to do with you.'

'But it isn't just my name, is it? She made friends with me.'

'But you said yourself that was just a strange coincidence, that you accidentally got one of her emails. It's not like she sought you out, is it?' Amanda says, her head cocked to one side, studying me like I'm being totally irrational.

She's right. Sally never came to find me; she didn't seek me out. But what I've not told Amanda is the fact that Sally had a newspaper article about me in her flat.

'What about Grace?' I ask. 'Did your contact find any record of her daughter, Grace?'

'No, I'm afraid not. To be honest, sweetheart, it sounds very much like this woman, whoever she is, might have been working up to some sort of con. Did she ever ask you for money?'

'No,' I say.

Amanda sighs. 'I imagine that's what she was working up to.'

I look down, trying to stop the hurt from blossoming over my face. I'm such an idiot.

'My friend said they think it's likely Sally – or whatever her name is – wrecked the flat herself and has now disappeared because she didn't want to face the consequences.' Amanda lets out a heavy sigh. 'I'm sorry, chick. I know it's not what you wanted to hear. But sometimes, you just can't help people like that.'

I look up at Amanda. It's the first time I've heard her be so disparaging about the sorts of people we come into contact with every day. She's never been one to make jokes about the unfortunate lifestyles we come across or get all high and mighty about the states people get themselves into, preferring to adopt an approach of 'there but for the grace of god'. Was that all an act to keep up her shinier-than-shiny reputation as Assistant Chief Constable? That's unfair of me. Amanda is human; she's only said what everyone else thinks most of the time.

'You've done the right thing now, Sal. The police have the report from Crimestoppers and they'll look into what's happened. The only connection this has with you is that she happens to have used the same name as you, that's all. It's not your problem.'

I nod, avoiding Amanda's gaze. She's right. Sally isn't who I thought she was. She's a liar and a fraud. She doesn't need – or want – to be saved by me.

'I should go,' I tell Amanda, suddenly wanting to be alone. 'Nath's on his way home.' I stand up and Amanda does too, following me out to the hallway.

'I know it must feel like a betrayal,' she says as she opens the door. 'But the best thing you can do is forget all about this woman. I think you've had a lucky escape, personally. It's likely she's picked your name out of thin air and then saw an opportunity to fleece you once you "made friends" online.' The way Amanda uses air quotes around this makes my cheeks burn. 'But either way, just be glad she's disappeared before she got anything from you. Now, give that son of mine a big kiss from me and I'll see you both on Sunday for a roast, will I?'

I tell Amanda she will and climb back into my car.

Sally Jones doesn't exist. There never was a woman living by the sea, happy with her daughter Grace by her side. For years, I've been emailing a person who doesn't really exist – and now that person is gone. I should be glad.

So why does it feel like I've lost a real friend?

–

Nath's gentle snores, which usually don't bother me at all, are driving me to the point of insanity tonight. I turn over and look at my Lumie on the bedside table, which helpfully lights up as I face it. One forty-four a.m. Jesus Christ. I've not got a single minute of sleep since we got into bed, while Nath was out as soon as his head hit the pillow.

We had a decent evening together, despite my skittishness when I finally got home after visiting his mum. Although I'd said I was happy with Chinese, he'd got a takeaway from the Thai on the corner that I love but can't usually justify the expense of, and laid it out on the table with candles and everything, grinning up at me as I walked through the door like a schoolboy. He is so easy to please; I'm so lucky in that respect. The simplest things make him beam, and he's always been the same way.

We talked about the wedding; I was pleased to be able to tell him I had spoken to his aunt, at least. It's coming around so soon now and I've got more left to do than I'd like to admit but at least that gave me a temporary reprieve to worry about something over dinner other than my obsessive thoughts about Sally.

But now the house is quiet – aside from Nathan's snoring – and no one is up to distract me, she's all I can think about. My mind keeps going back to Amanda's warning that I've caught a lucky escape that Sally's disappeared. Perhaps she's right. After all, she's not the type of person that would slot easily into mine and Nathan's life. Can I imagine a young woman like that joining us at Amanda's for Sunday lunch with Chloe and Matt? God, I really am a snob. It's a side of myself I've always known was there and never liked. Working for the police has only made it worse. But I defy anyone to deal with the people we do day in, and day out and not become weighed down by it all.

But Sally wasn't like them. Not the Sally I knew over email. She was funny, mostly about herself and the situations she and Grace got into, self-deprecating and clever. It's impossible to match up the woman from my emails with the picture of Sally's life I now have. I wish I knew how much of what she wrote was true and how much was complete garbage. I can't believe she completely made up Grace's existence – especially given what Tiff told me about her having a child taken from her. Something is going on here, I know it. Amanda might think it's a coincidence Sally picked my name to use, but I doubt she'd say that if she knew about the newspaper article I found in the wrecked flat.

'Sal?' Nath says sleepily from beside me, leaning across to switch on my bedside light as it gradually illuminates the room. 'Are you okay?'

'Can't sleep,' I say tightly.

'No, I can tell.'

I apologise, knowing I must be wriggling all over the place to wake him up. I do it when I can't sleep without even realising, that and clicking my tongue against the back of my teeth – an incredibly annoying habit Nath usually sleeps through, but not tonight.

'Come here,' he says, pulling me into his warm body. I lie with my back to him, his arms around me, and eventually drift off, safe in the security of being with someone who I know inside out.

CHAPTER FOURTEEN

The next morning, I wake with a headache to find a note from Nath that he had to head in early. It's barely gone six and I feel like I've had no sleep at all but know there's no point trying to get some extra hours in now that I'm awake; my mind is already busy with thoughts of Sally. Despite knowing this isn't her name, I have nothing else to call her. I dreamed of her last night; her and Grace. How can the little girl she's told me countless stories about over the years not exist? How does that fit with what Tiff told me? There must be a child – even if her name isn't Grace, even if she didn't live with Sally.

Downstairs, I make a strong coffee, then open my laptop at the kitchen table. I know Amanda said her friend had looked up Grace on the system and found no evidence of her existing, but I need to see this for myself. I go onto a birth registry website and start typing in the details I know about Sally's daughter.

> Name: Grace Jones
> Birth date: 2020
> Birth location: Weston-super-Mare

It's not much and the two most important factors – birth year and birth location – are guesses based on the fact I know she's five and Sally never mentioned the two of them living anywhere else. I hit search and take a sip of my coffee, anticipating the results.

> YOUR SEARCH HAS GIVEN ZERO HIGH QUALITY RESULTS. PLEASE PROVIDE MORE INFORMATION TO GET MORE ACCURATE SEARCH RESULTS.

I click through to another search page. This time it allows me to put in the mother's name and alternative birth years – I add in 2019 and 2021 and then a long list of names comes up. I work through them one by one, already with a feeling that none of these are right.

Once I've done the painstaking job of reviewing fifty-two records, it's clear none of these are my Grace. There's not a single entry with a Grace Jones born between 2019–2021, so either Sally didn't give birth in Weston, didn't give birth under the name Sally Jones, or didn't give birth at all.

After a quick google search I discover she could have chosen to give birth in Bristol, so I perform the same search with Bristol in place of Weston and get another seventy-six results. This time, I know exactly what I'm looking for and it takes only fifteen minutes to get through them all.

Amanda was right: officially, Sally and Grace Jones do not exist.

But Sally *does* exist. She had a flat in that name, and a job at that pub. That's somewhere I could try, I realise, before quickly googling it.

It closed down six months ago. For Christ's sake.

I sigh, frustrated. I told myself I needed to confirm what Amanda had told me, but deep down I was hoping that her friend got it wrong – that Sally and Grace were as real as they feel to me.

Amanda warned me to let this go, but how can I? She said this wasn't personal, that Sally using my name was nothing more than a coincidence, but it doesn't feel that way. I may not know the real Sally Jones, but she knows me. I've told this woman so much over the years. How can I stop caring about what happens to her? That's not who I am. If there is a chance I can help Sally, I'm not going to stop until I know she's safe.

I walk around the kitchen, imitating my dad from years gone by. I'd watch him like this at our kitchen table as he paced around the room, so deep in thought he'd barely notice I was there. I loved watching him work – the way he'd stroke his too-thick moustache like an old-fashioned detective from a crime novel, waiting for the answers to reveal themselves to him. I think what got to him most was that he could never solve the real mystery in our lives – where my mum had disappeared to.

Now, I do the same pacing motion with my feet as I try and channel some of my dad's energy. What would he do? Well, he'd never be in this situation to start with – but that sort of thinking doesn't help me. He was logical. He'd tell me to start with what I do know instead of focusing on all the things I don't.

There must be more to this story with her and Alex, her landlord. He's been forcing her into sex instead of collecting her rent payments – but she got an email that day to say she was going to be evicted. So, Alex's little scheme went wrong.

Before I can tell myself it's a bad idea, I phone his work line.

'Alex Harding, Layton and Harding, how can I help you?' His smarmy voice chirps down the line and curls my stomach.

'It's Sally Jones,' I say, my voice cold in comparison.

'What do you want?' he says quietly, as if he's trying to make sure no one else hears.

'So, I was wondering, how did it work with you and Sally? Did you pay the rent for her? Or did you just cook the books to make it look like there was money coming in where there wasn't?'

'What?' he snaps and I imagine his ratty face turning red in his shitty office.

'Well, you said you'd had this little arrangement for a while, right?'

'Yeah...'

'So, was it like a one slip and you're out kind of thing?'

'What the fuck are you talking about?' he growls.

'You sent Sally an email last week saying she was going to be evicted for not paying rent. Was that because you couldn't get her to sleep with you that month, or...?'

He sighs down the line and I hear him moving around, like he wants to get more privacy for this conversation.

'I told you,' he says, coming back on the line sounding slightly breathless. 'It wasn't an arrangement I could keep up forever. Dad started asking questions, looking at the books more closely...'

'And saw what?'

'That payments were missing on Sally's history.'

'So, you weren't paying her rent for her in exchange for the sex, you were just marking it as paid.'

'Why does that matter? She wasn't having to pay either way. Victimless crime!' he laughs, sickly. 'Dad would never have even noticed if he weren't so obsessed with getting these old shithole properties done up – suddenly needs all this cash, doesn't he?'

I drum my fingers on the table, uninterested in Layton & Harding's property portfolio growth plans. 'Are you sure your dad had no idea what was going on with you and Sally?'

There's a pause, a moment. 'Why would you say that?'

'I don't know. I'm just trying to work out why she has run away.'

Alex laughs. 'Well, it won't be my dad, if that's what you're thinking. He's five-foot-six and built like a pug – not exactly the most terrifying bloke on the planet. You're barking up the wrong tree, love.' He sniffs loudly down the phone and I wonder if Alex has something of a cocaine habit on top of his sex addiction.

'Do you not care at all that someone wrecked her flat and now she's disappeared?' I ask.

'Yes, I care. Of course I care,' he says, sincerely. Then adds, 'I care that I've got the police sniffing around the place and I've got a wrecked flat I can't rent out until I get rid of all her shit. Now, leave me the fuck alone.'

'You're a real piece of shit,' I say but the dial tone tells me he's already hung up.

So, the police are investigating now. Is that a good thing? I mindlessly pull up Layton & Harding's website, flicking through the available properties like I did last week when I first started worrying about Sally. On the last page I find details of the 'shithole properties' that Alex mentioned are being renovated.

> Situated amongst the beautiful bluebell woods on the outskirts of Weston-super-Mare, we are delighted to open for early interest registrations in our latest

> property empire: THE WOODLANDS. Expected to be completed in late 2026, register your interest now for a chance to be the proud owner of a property in what will be an exclusive, luxury development.

I scoff at the pictures of the 'luxury development'. Right now, it's a derelict building in the middle of nowhere.

Wait.

A derelict building in the middle of nowhere. The perfect hiding place.

I dial Alex's number again.

'Alright, I've had enough now – I'm going to block your number if you don't stop—'

'Those shithole properties you mentioned, the ones your dad is doing up. Did you ever take Sally to them?'

'Why?'

'Just tell me.'

'I might have,' he says in a way that tells me everything I need to know.

CHAPTER FIFTEEN

This is possibly one of the most stupid things I've done yet, and the list is already pretty impressive. But I'm here now – unbelievably I've made it back down to Weston without hitting a single bit of traffic on the M5 and the day is bright and quiet, not yet full of people going about their business.

I could have got this spectacularly wrong. I could be about to knock on the door of a drug dealer hiding out from the police or a sex trafficker or… oh god, just stop. As soon as Alex told me he'd brought Sally to the derelict building, I knew I was going to come here. The moment I got the email saying she was going to be evicted, I knew I'd do anything to help her. I can't explain the overpowering need to save this woman, but I can't deny it either. For all her lies, we're connected somehow. It's as simple as that.

In a routine now familiar to me, I drive along the seafront watching the waves steadily coming in and going back out again against the pebbly mud-like beach. Google Maps tells me to keep going, further along the seafront than I'm used to. I drive past an old burned-out pier and vaguely remember visiting this as a child, my dad making up ghost stories about the victims that still haunted the pier to this day, and my mum telling him to stop it or I'd get nightmares. The pier is spooky, but I know no one died here. Probably bored kids messing around and it got out of hand. Isn't that how most horrific things happen in life? A simple misstep gone too far.

I follow the road as it curves around the clifftops taking me further away from the town centre, further from civilisation. As I turn away from the seafront, into what looks like a wreckage yard, I wonder if my directions are completely wrong, but Google Maps confidently declares I have arrived. I turn off the engine and step out of the car, pulling my black puffa coat around me despite the sunshine. There's less wind than I've grown used to down here; we're protected by a thick wall of trees that are out of sync with the ugly buildings on the yard.

I try to imagine Grace here; a little five-year-old girl who loves pirates as much as she loves princess dresses, but I can't. Perhaps that aligns with the fact that the little girl I've heard so much about over the years doesn't actually exist. Grace aside, I can't imagine anyone living here at all. Town is only two or three miles down the road, but you'd never walk it; the hill from the yard to the main road would be enough to put you off but you'd never take your life in your hands by walking along the cliff edge either. But perhaps that's exactly the point of this location; no one is coming here by accident.

I follow the barely-there path made by workers' boots through the mud and walk until I reach an old caravan, giving it a wide berth as I imagine *The Hills Have Eyes* characters jumping out at me at any moment. The door to the caravan hangs off precariously, like it's been a long time since anyone used it, but still, you never know. Moving forward, I follow the path left into the woods, counting my steps to calm myself.

As I'm crunching over the buried gravel path, my phone pings, scaring the life out of me.

'Jesus Christ,' I whisper to myself as I pull it from my pocket, looking around as if the sound might wake up the ghouls hiding in the trees. When I'm certain nothing is lurking within, I read the message on the screen. It's from Nath.

> Been pulled onto a night shift again today so don't expect me back. In ten days we'll be getting married! Love you x

Ten days feels inexplicably soon. I try to imagine slipping into my beautiful wedding dress that I chose with Amanda in a little boutique in one of the local villages, but I can't. We were both so delighted to find my dress so quickly and it now hangs, full of promise, in her spare bedroom wardrobe. I walk forward through the woods and try to picture myself walking down the aisle to meet Nathan at the end. The idea should calm me but instead I get that feeling again, that I shouldn't be here at all, that I should be at home planning for what everyone keeps telling me is the 'most important day of my life'. I should turn around, get back in my car and drive away. But then up ahead, there's movement in the trees. Is it her? Is it Sally?

I pick up my pace as I come off the track and into the trees and scan my surroundings, trying to find the source of movement but there's nothing. Perhaps it was a fox. Taking a deep breath, I push forward. Surely it must be close by?

There it is. The dilapidated building I looked at on Layton & Harding's website this morning. No wonder Alex's dad is worried about financing this; it's in such a state of disrepair it's hard to believe anyone will ever live here again. There's no sign of work being done either, so any renovations they've got planned can't have started yet.

Now I'm here, it feels unlikely this is where Sally is hiding. Yet a few hours ago, I was sure it was the key that would unlock the whole puzzle of Sally Jones. I hate to imagine why Alex would ever bring her here, the whole place is so grimy. In the distance, you can hear the seagulls squawking, but apart from that, it's impossible to know that you're so near the sea. The tall trees overhead block out most of the light and I wonder if Alex's dad will arrange for them to be felled so his luxury renters can enjoy unencumbered views of the crashing waves.

I don't have a key for the building, but I doubt I'll need it. On the mucky path, there is a set of footprints. I stand my own feet in them – the same size. Are they hers? Is Sally in this building, hiding from something, someone? I take a step forward, pushing open what is left of a front door.

I walk into the dark, cold building. What if it's not her? There could be anyone hiding away in this abandoned building in the woods. Anyone at all. I come back out, gathering myself.

I stand outside the building and look back up at it. There are three floors at least and I imagine myself going through every single room, floor by floor, the sound of my heart banging in my ear drums. Overhead, a bird – a red kite, I think – circles as if I'm his prey for the day and I shiver. This at least pushes me forward and I walk back to the door then wait and listen, as if I might hear Sally calling out to Grace and the little girl laughing in reply. Despite everything I've learned about Sally, it's this scene I want to find most of all. I want Grace to be safe. More than anything, I want Grace to be real.

I'm inside. The smell of damp fills my nostrils and it's dark – unnervingly so. I get out my phone and flick on the torch.

As the corridor lights up ahead of me, it's clear this building was never residential, but was once something else entirely. A hospital, perhaps? The ceiling is filthy but looks like it might have once been white; large strip lights hang above me as I walk, avoiding parts where bits of rotten and likely asbestos-filled ceiling have long ago fallen down. I listen for any signs of life around me, but can only hear the pounding of my own heart and the occasional scratch and rustle of small animals. This would be the perfect place for rats, I suppose. As if thinking of them has conjured them up, I see a flash of a long tail darting between the rubble.

I reach a series of doors all along the corridor, some hanging open that I quickly peer my head around and discount – they show near-empty rooms, wires coming out of the walls with high-up windows so filthy the sun barely penetrates. I keep walking, my feet leaving footsteps in the dust.

I reach the end of the corridor, the weight of my own stupidity heavy on my shoulders. There's no one here. Why did I think Sally would be hiding out in an abandoned building? Another case of my instincts being off.

But then I hear it – a cough.

I whip around, looking back down the corridor I've just come from – but I know I checked every single door, there was nothing and no one behind any of them. The cough must have come from outside. I slip out through what was once the back door and is now a hole half-filled in by rubble. Outside, there's another building. It's small, squat and square. Like it might have once been a caretaker's house or similar. But the key thing about it is that it doesn't look abandoned. Not at all.

I step forward through the rubble, moving further away from the abandoned building, a sense of excitement building in my limbs as I get closer and closer to the small house. There's that sound again. A cough followed by a gentle thud of footsteps inside.

There is *definitely* someone here.

I lift my hand to knock on the dark brown door as a large bird swoops down from above and lands in the nearby wood, a screeching sound emitting from the trees, setting my teeth on edge. I knock. Loudly. I wait. I take deep breaths, I count backwards from ten, then knock again. There is someone in this house, I know it.

I hear footsteps on the other side of the door but they're hesitant and I imagine whoever is standing on the other side is pressing their eye up to the spy hole, though the muck blurring the glass won't allow them much of a view.

'Sally?' I say, my voice hopeful and too loud in these woods. 'Is that you? It's me. Sally Jones. I'm here to help. Can you open the door?' I sense the person on the other side of the door and listen as a chain is unbolted, the lock undone and then, finally, slowly the door opens.

'Oh my god,' I hear myself say, as if from a distance, as the woman I've been looking for appears before me. 'It's *you.*'

PART TWO

CHAPTER SIXTEEN

NATALIE

Why the fuck is Sally fucking Jones at my door? This is a disaster. I try to slam the door shut but she's too quick, sticking her muddy boot in the gap and stopping me.

'Piss off!' I shout. I sound like a proper teenager but so what? It's not like I need to keep up this '*I'm just like you*' act for her any more. Why did I even open the bloody door in the first place? It was a spur of the moment decision – one of my most fucking stupid – because I'm lonely as hell and the sound of another person just bloody undid me for a second there. I won't make that mistake again.

'Please,' she says in the posh voice I'd imagined from her emails. 'Let me in or I'll call the police.'

I try and shove the door closed again but she won't move her stupid foot so after a bit of bashing, I give up and let it swing open on its hinges, and we look at each other properly for the first time.

'Call *yourself*, you mean?' I say with my arms crossed as we stand face to face.

Her brows furrow and she takes a step back further into the mud, which makes her frown even deeper and I try not to laugh. This stupid bloody woman.

'Are you who I think you are?' she asks like she's trying to be threatening but there's uncertainty in her eyes, maybe even fear. I should be pleased she thinks I'm a threat but it's pretty horrible actually, having someone who used to be a friend look at you like this when they think they've discovered the real you.

'Don't pretend to be shocked,' I say, turning to walk back into the house as if it's going to be that simple to walk away. 'You've always known who I was.'

'It's really you, isn't it?' she says from behind. 'I recognise your face from your mugshot. And your voice. I'd remember that voice anywhere.'

I keep my back turned, her words like little darts landing in my spine as I take a step away from her, away from this. Is she having me on? Or is my identity actually a shock to her?

'You're Natalie Pierce,' she says – sure of herself now – and I stop.

It's been five years since I've heard that name and part of me wants to turn around and shout 'Yes! That's me!' as if finally, someone knows the truth. But the Natalie Pierce she knows and the woman I want to reclaim are two totally different people.

'You murdered Shane Blackwell,' she says, and I turn, coming face to face with her. I start shaking my head. 'You abandoned your baby! You left her to—'

'That's a lie!' I shout, pushing her back so her warm breath is away from my skin and I can't taste the hate in her words. She stumbles backwards but then rights herself against the wall. My chest pounds and I know it was the wrong thing to do – to push her – but how can she still think those things about me? How could anyone?

'This is completely insane,' Sally says, pulling her phone from her pocket but I'm too quick – I grab it from her hands. She yelps like a hurt puppy.

'For god's sake,' I say. 'Just stop it. Alright? You've messed things up enough for me already. Can you just stop?'

'*I've* messed things up for *you*?' she says with a tone that says *how could someone like her possibly ever inconvenience someone like me?* and I want to push her again, but I don't. Instead, I walk into the kitchen – if you can call it that – and place her phone calmly beside me next to the rusty kettle.

I'm not going to act like the exact type of person she expects me to be. Isn't that what the last five years have been about? Proving to myself – before I can prove to everyone else – that I'm better than that. That I'm worth something. Despite what the rest of the world might think.

'What do you want, Sally?' I say as I watch her hover awkwardly in the kitchen doorway, trying to work out if her shock is an act. If it is, she's a bloody good actress.

She can't step fully into the room because if she does, we'd basically be within touching distance and I can tell by the way her nostrils flare that she doesn't want to touch my skin. Like the grease of me would stain her perfectly white t-shirt.

'I want…' she says then stops like she doesn't know. Was I wrong? Does Sally genuinely have no idea what she's got herself involved in?

'I want to help you,' she says.

That's… interesting. 'Help *me*?'

She sighs. 'I don't know what's going on, but I know you're in trouble.'

I give a funny little laugh because I can't exactly deny that statement, can I? I'm hiding out in a shithole in the woods.

'What's that got to do with you though?' I ask, testing the waters.

She moves from foot to foot. 'Well... I don't know!' she exclaims, throwing her hands up. 'I don't know what the hell is going on, do I? I thought I was coming here to help a friend and instead I find *you*,' she says the last word with disgust. 'Why did you steal my name?'

I shake my head. As if that's the most important thing right now!

'I didn't *steal* it,' I say, ducking my gaze away from hers to the dirty kitchen floor. 'I borrowed it.'

'My god,' Sally says. 'This is unreal. Natalie Pierce. All this time, I've been talking to Natalie bloody Pierce.'

I look up, imagining what that name means to her. What she remembers when she says it. And I'm sure now, that this is brand-new information to her. All this time, she's really had no idea.

'I helped you that night,' she says, almost in a whisper.
'I know.'

'Why did you... How could you?' She steps towards me, then steps back, her face wary and it's like watching the memory play out in her mind.

What will she have read about that night? Probably more than I have. I only saw the newspaper reports and they never give all the gory details, do they? The sound he made when his life was ended. The blood that pooled at the back of his bashed-in head. But Sally would have read all the police reports. And she wouldn't care about a man like that – no one cares about men like Shane. It's not his murder that turns her stomach when she looks at me.

'I didn't do what they say I did,' I tell her, expecting some level of understanding, some level of trust after the years of emails we've shared, but instead she scoffs before she's even had time to think about my words, barely hearing them at all.

'Right,' she says and it's like her earlier offer of help never happened at all.

'I didn't!' I shout, making her flinch. Good. Be scared. If I can't make her like me, maybe this will have to do.

'Why did you contact me?' she asks. 'That first email. What did you want?'

I grit my teeth. As if *this* is what she's stuck on! 'I didn't. You contacted me.'

'Only because I got a rogue email meant for you—'

'That's hardly my fault, is it?'

'Oh come on!' she says. 'You planned it. You stole my name and then you used that to send me an "accidental" email knowing I'd forward it on to you.'

I curl my face up. 'Are you serious? How would I even…' I let the sentence hang, as if I don't fill the gaps, she'll realise how stupid she sounds. Like I'd got this mad master plan to get in touch with her! 'I never wanted to speak to you again,' I say when she doesn't say anything. 'I just needed a new name and yours… Well, I thought it was a decent name.'

She eyes me suspiciously. 'A "decent" name?'

'Yeah,' I say, sniffing. 'Like the name of someone good. I wanted to be… someone good.' I let the truth hang between us in this shitty, filthy kitchen, a desperate move by a desperate person.

When I saw the newspaper article about Sally not long after I'd gone into hiding, I thought it was the perfect name to take as my own. She was the woman I'd spoken to on the phone that night, she was the one who'd said the words to me that I've repeated to myself over and over since then, the only thing that's kept me going. She was also the person I had to thank for keeping my little girl safe. At least, that's what I thought. Now, I'm not so sure about any of it.

'You knew who I was though, didn't you?' Sally asks me and I nod. 'We spoke. That night.' I nod again. 'I believed you,' she says, and I search her face. If I look hard enough it looks like hurt. But why would *she* be hurt? 'You said you just wanted to keep your baby safe. You said… You begged me for help.'

I dip my head, remembering with a punch our conversation that night.

'You told me I was a good mum,' I say and despite myself, tears threaten as the back of my throat burns.

Sally's eyes fill with matching tears, and we stand, face to face, quiet for the first time since she got here.

'I know what you're thinking,' I say. 'I saw what they wrote about me in the papers after. They said I was a bad mum, that I was hurting my little girl, but I wasn't – I never did any of it. I was a good mum; you were right about me. I might not have done a lot of good things in my life, but I know I was a good mum.'

'Natalie, they found your flat in a total state. Unfit for a child. Your baby's cot… Her cot was drenched in alcohol and you'd no food in the cupboards.' She looks at me with disgust, taking a step back.

I shake my head fiercely, stepping forward to close the gap between us and she flinches. 'That's not true! That's not how I left my flat. I swear to you. I would never, ever have treated Grace like that. I loved her. She was my everything.'

'Grace…' Sally says, her eyes rounding in recognition as if she's only just cottoned on that the little girl in my emails is the baby she 'saved' that night. 'Grace was—'

'Grace *is*,' I say with more force than I mean to, 'my daughter.'

Sally shakes her head and looks over her shoulder back to the door, as if she's thinking about walking straight out of this shithole. I can't let her. Not now. She knows too much; it's too dangerous.

'Sally,' I snap, grabbing her by the wrist so she looks at me. 'Please, please listen to me. Do you promise you're not involved in this?'

She pulls her wrist from my grasp and looks at me like I'm insane.

'Involved in what?' she asks. 'I have no idea what "this" is!' She throws her hands up and I flinch as her elbow grazes my arm.

The dingy kitchen is too small for both of us. We can't talk like this; I open the door to my right, to the living room-cum-bedroom and tug on her sleeve to follow me in. At first, she resists but then lets herself be dragged through.

It's just as grimy as the kitchen in here, with the room's thick curtains blocking out any natural light – not that if I opened them it'd be any better because the windows are caked in years of filth and bird shit – but there is at least a place for us both to sit. I pull the thin duvet off the sofa bed and point for her to sit beside me. She inspects the stained blue sofa cushions as if they might give her some sort of disease but eventually perches herself down next to me.

'Look, what happened that night… It was messed up,' I say, and she laughs as if this is the understatement of the year. 'But I promise you, I'm not the one who killed Shane. I was set up—'

'By who?' she says and I can already tell she doesn't believe me.

I shake my head. I can't tell her the truth, not until I'm sure she really isn't involved. 'It's complicated…'

'You know the police are still looking for you,' she says, all prim and proper. 'One call and they'd be here in a second.'

'Then why haven't you called them?' I snap.

She shifts uncomfortably and shrugs. 'Honestly?' she sighs, loudly. 'I don't know. I want to understand what the hell is going on, for a start.' She looks at me, her eyes narrowed and unsure. 'I can't believe it's *you*. All this time it's been you. Your phone call that night changed the course of my whole career,' she says.

'I know,' I say with a smirk. 'Got you a big fat promotion, didn't it? And a fancy award.'

'So you *did* know who I was when you stole my name.'

I shrug, like it's no big deal, but I feel my cheeks reddening. 'I read about it in the paper. I remembered your name, *Sally*, from when I called the police that night.' I pause as the sound of the steady drip from the kitchen tap marks the seconds between my words. 'You had a kind voice, that's what I thought when I first spoke to you.' I look up, she looks freaked out, so I quickly add, 'But I swear, I didn't take your name because of anything except needing one and liking it. I promise.'

She studies me, unsure whether to trust me and I don't blame her.

'After that night,' I say, working out as I go how much to tell her. 'I had to disappear. I was planning on lying low for a few weeks, until it all died down, then coming forward and telling the truth—'

'Why didn't you? If you're so innocent, why didn't you tell the police that?'

I shake my head. 'It's not that simple, is it? I found something out and by then it was too late to go back. Everyone thought I was a murderer and a…' I take in a big breath because I hate thinking this, let alone saying it. 'Child abuser.'

CHAPTER SEVENTEEN

People are a lot more comfortable with a murderer than a child abuser, you know. I've had a lot of time to think about it and the thing is, with a murderer, they can reform. You see it all the time in the papers, the way they bang on about a murderer being pushed to the edge. Even when the crime is proper horrific, like. People get that you can be pushed into killing someone – like it might happen to them.

But if you've hurt a child, abandoned a baby? That's it. No reform for you. And I get it, I do. The thing is though, I'm not either of those things. I didn't kill Shane Blackwell and I definitely didn't hurt my Gracie – not once. I loved my little girl more than anyone can ever imagine. Being set up for murder was one thing, but being framed as a bad mum who abandoned my own baby was worse.

'You know where they took her, don't you?' I ask Sally now.

I watch her pretty face – because now that I look at her, she really is pretty. Kind of mousy, I guess. But she has that clean look about her. You know, like she has a 'skin care routine' and probably double shampoos her hair like they always tell you to do on TikTok. We don't look alike at all. I thought we might, you know. Even though I'd seen that photo of her in the paper from years ago, over the years of emailing I've let myself forget we aren't anything like each other in real life. She's healthy. Even as she glares at me, her skin glows. She probably eats avocados. My skin's dull and grey and I'm scrawny fat – the worst kind. I'm distracting myself comparing the two of us while I wait for Sally to reply, even though I know she isn't going to, because more than anything in this whole stupid world I want her to reply. To tell me where my little girl is now.

'Please,' I say as she bites down on her pink lips. 'Tell me. Is she okay?'

Sally sighs and fiddles with the ends of her glossy brown hair. 'What do you want me to say?' she asks, sounding tired, like this is wearing her out. 'I want to help you, really, I do. But I'm not involving your child in this—'

'Grace, her name is Grace.'

Sally shakes her head. 'No, it's not. Not any more,' she says and it's like a gut punch.

'They changed her *name*?'

Red anger curls up my neck. They stole my baby and then they took away the only thing I was ever able to give her. I spring up from the sofa and pace the small room, my body burning with rage.

'How could they change her—'

'No one knew her name, Natalie,' Sally says in this stupid calm voice. 'You never even registered the birth. How could anyone know you called her Grace?'

I stop pacing and look at her.

I never registered the birth. Is that right? It doesn't sound right but things were so bad when Grace was first born maybe I forgot. It sounds like the sort of official crap I might have forgotten to do. My anger starts to fade a little at these faceless people who took my girl and instead shame creeps in. It's *my* fault my baby girl doesn't even know her own name.

Sally clears her throat nervously, the same way she did that night on the other end of the phone.

And then I remember.

'I told you her name,' I say and Sally's cheeks colour.

'You didn't.'

'I did,' I say, sure of myself now. 'I told you her name on the phone. I said Grace. You should have written it down. Remembered it. But I guess you didn't think it was important enough.'

Sally sighs and rubs at her eyes, leaving specks of black mascara on her freckled cheeks. 'If you told me, I'm sorry. I don't remember that. There was a lot going on. If I'd known her name was Grace, I'd have made sure she was able to keep it. I promise.'

I nod, the wind taken out of my sails a bit by her genuine apology in the face of my shame that I didn't even bother to register my little girl's birth. What sort of mother does that?

'What's her name now?' I ask, my voice small and pleading, but Sally shakes her head.

'I'll tell you,' she says. 'But you have to work with me. We have to trust each other first. Okay?'

I chew at the rough skin on my thumbnail as I toy with the idea of trusting Sally. She can get me closer to Grace, which is all I've ever wanted, even if I've not let myself believe it could ever happen. But will Sally actually help me or will she tell me what I need to hear then mug me off and hand me to the very people I've been running from?

'What do you want with me? I don't get it,' I say.

Sally laughs, a nervous almost snort. 'Honestly, I have no bloody idea.'

We've both calmed down a bit now and Sally is in the kitchen making us tea. She walks back in with two chipped white mugs and a frown on her face.

'Your milk's off,' she says, and I shrug. 'And these were the only mugs I could find.' She hands one to me.

'Yeah well it's not exactly the Hilton in here, is it?' I say.

Sally sighs and sits back down on the sofa bed. 'No, it's not.'

I look around the room and try to see it as she does. It's pretty fucking dire – even from my point of view – so god knows what she's thinking.

'I had my flat pretty nice you know. Before it got all messed up,' I say. 'I was making it a home.'

She nods but it's obvious she doesn't believe me. I know what she thinks. People like me live in shitholes. We have fag-burned sofas but massive TVs. Our banned-bullies sleep in the garden tied to the washing line. We're scum, basically. And I can't blame her for thinking that because half of the time, it's true. I am scum. But I'm trying so hard not to be.

'Who did that?' she asks. 'To your flat. Who wrecked it like that?'

I shake my head, blowing on my coffee so I don't have to meet her eye as I remember the way they turned my home into a fucking bombsite.

'Looked like they were pretty angry at you,' she adds but I still say nothing because it's not time – not yet.

'I met your landlord,' Sally adds, and I curl my nose and look at her.

'My landlord?'

'Alex.'

'Oh,' I say. 'Him.'

'Is it him you're running from? Is that why you're here?' Sally's freckled face looks so hopeful and innocent that I can't help but snort out a laugh.

'Alex? You think I'm running from Alex?' I laugh a little more, but it only makes Sally look more confused, so I stop.

'I know what he was doing to you,' she says, and I stop smiling then as the familiar creep of shame slowly covers my body.

I sniff, shrugging my shoulders to shake it off. 'Yeah, well. Didn't have a lot of options if I wanted to keep a roof over my head, you know.'

She throws me a sad smile as if she understands but I know that she doesn't. How could someone like her ever know what it's like for someone like me? I bet she's got family all over – a mum and dad who still look out for her even though she's an adult. I bet they call in on the weekend and bring her flowers – them multi-coloured tulips – to brighten up her already sunny living room. She won't get men like Alex; she won't get women like me.

'I understand, you know,' she says. 'We see things like that all the time at work—'

'Seeing it and living it is pretty different,' I snap, and she nods again.

'I know.'

'Anyway,' I say, wanting to change the subject from Alex. 'It's nothing to do with him. He's just a... he's a cockroach. Something you have to deal with because it just won't die, but it's never going to kill you.'

I watch Sally battling with her inner saviour, desperate to tell me what Alex does is bad; how I should report him to the police – all that shit people like her always say because they have absolutely no clue what it's actually like to be in the position I'm in. They can't imagine it. That's the problem. But they still think they can solve it.

'So, if it's not Alex you're running from, what is it?' she asks. 'This time, I mean…'

I laugh a bit because it is a little funny how she needs to add that bit of explanation because I have run away twice, she's right. I ran away from my life as Natalie Pierce and now I'm running away from my life as Sally Jones. I wouldn't have, you know. If the money had kept coming, I'd have stayed hidden away as Sally Jones forever. But it stopped, and then Alex started prowling around, and then I found out more about the real Sally Jones and they wrecked my bloody flat!

'The past caught up with me,' I say and immediately feel like a twat because I sound like some old bloke in a crappy detective film. I look away from her, embarrassed, and start picking at my nails again.

'Which bit of your past?' she persists.

'All of it,' I say leaning back on the sofa and stretching because I don't know what to tell her. I've always known who Sally is – that she was the person who took my call that night. But I didn't know the whole truth about her until recently – who she's involved with – and when I found out, I had to repaint our entire relationship into what it really was: a scam. I thought she was setting me up, or at least keeping an eye on me for *them*. But now it seems like I was wrong. She's totally clueless about what she's involved in here and I don't know what to do with that.

She sits silently for a minute and I close my eyes, leaning my head back against the sofa cushions because

I'm tired. I'm so, so tired. I don't sleep well here. I don't sleep well anywhere, haven't since that night, but it's worse in this creepy little house in the woods. I was totally mad to come here. I had no plans for what to do next, no idea how I'd get back on my feet this time. Nothing. Then in Sally strolls, a hand reaching out to connect me back to the world – back to the thing I want most of all.

'Natalie,' she says, inching towards me on the sofa like she can hear my thoughts. 'If you keep running, how are you ever going to find Grace? Let me help you, please. I want to help.'

I sigh. Sally thinks she's my friend; she thinks she wants to help me. And I want to let her. But what she doesn't realise is that to fix my life, I'm going to have to completely and utterly wreck hers.

CHAPTER EIGHTEEN

Sally's trying to convince me to go to the police now. She says whatever I'm running from, it can't be worth spending my life hiding away in shitholes like this. I do try not to laugh, honestly, I do. But it spills out like it used to do when I was at school and then the teacher would give me the look Sally is giving me now. That screwed-up face that says 'what is wrong with you? Why are you like this?' and it only makes me laugh harder because I don't fucking know.

'Natalie,' Sally says, placing her hand on my arm. 'I'm really worried about you.'

'You don't even know me!' I say as the laughter stops.

'Yes, I do. We've been emailing for years.'

I roll my eyes. 'You've been emailing a made-up person. *Sally Jones* who lives in a beautiful terraced house with her gorgeous daughter Grace.' My voice cracks on her name. 'Sally Jones, whose biggest problem is that some random woman keeps getting her emails by mistake.'

Sally watches me and I expect her to turn away, to come to the realisation on her own that everything she's ever learned about me is a lie. A trick to keep her close. But she doesn't turn away, she just looks. And looks. It's unnerving. Like she's looking right into my head.

'I know a lot of what you told me wasn't true,' she says. 'But I don't think all of it was a lie. I think we were friends. We *are* friends.'

I start to snap something spiky back but the words won't come and instead I tear up like a right tit. If only she knew.

'I want to trust you,' I whisper because saying the words aloud, trying to sound vulnerable, is the opposite of everything I'm used to.

'Then trust me,' she says, the hand squeezing my arm. 'Let's get out of here. Please, Natalie. I'll take you somewhere safe. And when you're ready we can talk to the police—'

'No police,' I snap, standing up and shaking off her touch. 'You can't trust them. *We* can't trust them.'

Sally blows out the air from her cheeks like she's battling with the idea the people she works for – the person she is – can't be trusted. That they're the bad guys. She trains her eyes back on me and I think she's going to tell me it's her way or nothing, but she doesn't.

'Okay,' she says. 'But you have to tell me something, Natalie. Tell me anything.'

My heart bashes against my ribs. What can I admit to Sally that won't completely fuck up any hope I have of her helping me get Grace back?

'All I can say is that it's in someone's interests that I stay hidden, and when that someone found out I might be getting sick of hiding away, they...'

'Is that what happened to your flat? Someone trashed it because they thought you were going to come out into the open?'

I nod.

Sally digests this. 'Why do they want you to stay hidden? Is it someone who knows about Shane? Is it connected to his drugs gang?'

I sigh. 'Yes,' I say, confidently – because it's not wholly untrue. 'Shane is… Shane was the reason that my whole life got fucked up.'

Sally tilts her head to the side. She won't understand, how could she? She'll know of people like me, of course she will. I bet she speaks to hundreds of us every single day in her job. But speaking to someone and understanding them are totally different. If you've not grown up like I have, you won't understand how someone like Shane can walk into your house and seem like a saviour rather than the devil.

I was fourteen when I met him. I felt like an adult. I would tell anyone that listened that I was 'wise beyond my years', something I'd heard one of the Olsen twins called on that TV show they used to have. I used to love that show. I can still remember the theme tune; I'd watch it while Mum hacked up her lungs in the bedroom next door, turning the volume up to cover the sound. Mum was pretty much on her deathbed by the time Shane set his sights on me. No one even knew she was sick. She'd made me promise I wouldn't say anything because if I did, the social would come and take me away, and she was terrified to die alone.

So, I never did tell any adults what was going on; none except Shane. He'd come around with flowers he'd got from the Nisa around the corner but it felt like something out of one of the romantic films I used to watch set in New York. I'd pretend he'd picked them up from a proper flower shop, choosing each pretty colour especially for me. My mum properly loved him. She used to say she was so happy I'd found someone to look after me when she was gone. I'm glad she died before she understood that she'd handed me over to a wolf.

'Is that your stomach?' Sally asks, jolting me out of my memories.

I hold my hand against my belly to stop the rumbling. It's been like this for days now and I've worked out the best way to get it to stop is to clamp down as hard as I can against myself, as if the pressure is causing the pain instead of my stomach eating itself because I can't give it any food.

'When was the last time you ate?'

I shrug, not because I'm being a twat about it but because I genuinely don't know.

'You must be starving.'

I shrug again. 'I'm alright.'

Sally stands up. 'Come on. Let's go and get some fish and chips.'

My stomach gurgles loudly and the edges of my mouth literally water at the memory of the salty tang of vinegar on chips so hot they burn my lips, but I shake my head. 'I can't. Someone might see me—'

'No one will see. You can go in the back of the car and lie down like we're in some silly spy movie if you want. I promise. No one will see.'

I eye her cautiously. The idea of chips is enough to draw me out of this hell hole to be completely honest.

'If someone sees me…' I say, fear gripping me around the ribs and scaring my hunger away. 'No,' I say shaking my head. 'It's not worth it.'

'Natalie,' Sally says, reaching out for my hand and holding it in hers. I start to pull away but then give in because it's nice to be held by someone, even if they are a stranger. Well, kind of. 'I promise, I won't let anything bad happen to you.'

So many people have said that to me that it's hard to believe, but there's something in Sally's face that makes me want to trust her, it's like she really does want to help me. A pang of guilt hits me because if she knew what she was doing – what she's unravelling – she'd run a million miles.

'If you want to find Grace,' she goes on. 'You're going to need to leave here sometime, right?'

That's what does it for me. Because she's right. There's no world in which I find my little girl by hiding out in a shithole in the woods. I'm going to need to take some action and maybe Sally's right. Maybe I can trust her. Even if she can't trust me.

CHAPTER NINETEEN

We walk out of the house hand in hand and it's weird to let Sally lead me like this, like she's the one in charge. I suppose that's how she's used to feeling with me. Before I knew what I know about her now, I thought she was perfect; I was happy to be led. I loved emailing her. On my darkest days it was the only thing that kept me going; this link to the outside world, to normal people with lives and families that didn't exist in my shitty tower block.

I had Layla; she was a good friend. But her life was just like mine. Covered in shit. She'd been born in that tower block and she said she'd die in it, too. She's only twenty-five – like me – but has already given up hope of things ever getting better. Before Sally, I felt like that too. That I'd never escape.

But then Sally's name appeared in my inbox and her friendly emails became a reminder that people did live lives outside of my small, nasty world. It seemed like fate that the woman who told me I was a good mum, the woman who believed in me that night, had accidentally got my email and found me like that. But perhaps I was willing it into being by taking her name in the first place…

I loved the life I lived through our emails. It was an escape from the everyday hell that was my life. Instead of an unemployed, unloved woman reliant on the 'kindnesses' of men to get by, I became a mother with a little girl who loved me. I became a real person; that was how Sally saw me, and I loved it. I was a real person, a person just like her. Things might have stayed that way forever, me living out my fake life, if I hadn't found out the truth about Sally's real one.

We reach Sally's car and I'm surprised it's not one of those huge stupid Range Rovers all them posh women drive but instead a grey hatchback with rubbish strewn on the backseat.

'Sorry about the mess,' she says although she doesn't seem particularly sorry and I think maybe she isn't quite as a prissy as I've pegged her to be. It's hard not to like her, but I can't let myself think of her as a friend. It'll make wrecking her life so much harder.

I get in the car and she slides into the driver seat. The brightness of the day makes me squint; I'm not used to sunlight after over a week hiding behind those rank curtains. I wind the window right down to let the fresh air onto my skin and close my eyes as the sun tickles my face.

I know it's stupid, but I really want to trust Sally. Maybe she isn't the same as all the rest of them. She seems so naïve – like she really doesn't know the truth of what happened back then. If she did know, would she really want to help me? I watch her from the corner of my eye. She helped me that night. Maybe she'll help me now. For the first time in years, I feel something. I think it's hope.

'Shit.'

I watch as Sally's eyes grow wide and panicked at whatever she's seen up ahead, then look up to see for myself.

There's a car. Someone is here.

'You called someone,' I say, all hope flying out the window as fast as it arrived.

'No,' Sally says, her head shaking back and forth chaotically. 'I didn't, I promise I—'

'You fucking liar. How could you do this to me?' I try to open my car door but Sally's too quick, she snaps the locks shut and I'm trapped.

I watch as the other car slowly crawls through the woods in the distance towards us and I see everything going to absolute shit right in front of my eyes. How fucking stupid I've been. Is it the police? Am I about to lose absolutely everything?

'Let me out!' I scream at Sally, my hands flailing around me, smashing against the locked car door.

'Stop it! Natalie, I didn't call anyone, I swear I didn't—'

But I can barely hear her words through the thunder of my anger; every second we sit here they get closer and my chances of ever seeing Grace again get further away. The car edges nearer to us then stops. It's still too far away for me to see the driver's face but once they get out, it will take less than a minute for them to reach us on foot. Then, that'll be it.

'Drive then,' I say, grabbing Sally's hand. 'Show me I can trust you and drive!'

She looks at me, the fire in my eyes burning back in hers. She tightens her grip on the steering wheel, then slams her foot down on the accelerator.

The engine roars, dust flying around the car and I'm half terrified, half elated. A mad giggle escapes my throat as we speed towards the stopped car; Sally accelerates as she looks over her shoulder to turn right and come out of the woods and I start to think we might get away after all – and then it happens.

The car door opens, and a person steps out.
Right in front of us.
The body flies over the roof, cracking the windscreen and we scream.
Shit, shit, shit.

CHAPTER TWENTY

The body on the ground is still, lifeless. In the wing mirror, I see an arm twitching and release my breath. Maybe it's okay. Maybe it isn't that bad... But there's no time to think about what just happened. We need to get out of here now. I turn to Sally but she's gone into shock, her hands clasped against the wheel, her big eyes wide and staring forward at nothing.

'Sally,' I say, grabbing at the steering wheel. 'You need to go.'

She doesn't move. Then looks at me, her face white, her eyes empty.

'Go,' I shout, smacking her arm, hard. 'Jesus Christ, Sally, we need to get out of here. Now!'

The smack shakes her out of the shock and once again she slams her foot down on the accelerator. In a minute we're back on the road and I'm directing her like I'm the one in control and we drive along the main road out and out until Weston can't be seen any more. I keep expecting to hear the scream of sirens behind us but they never come and when we pull over into a pub car park, the silence is deafening.

'I don't understand…' Sally says, hands still on the steering wheel in the ten and two position. 'I don't understand what just happened.'

'He walked out in front of you,' I tell her. 'It wasn't your fault.'

She searches my face, confused like a toddler – she's heard the words I've said but it's like she doesn't understand their meaning.

'What?' she asks, her voice small. 'What do you mean…'

'Alex,' I say, his name like grit in my mouth. 'He jumped out of the car and stepped right in front of us.'

'Alex?'

'I'm going to check your car,' I tell her, unstrapping my seatbelt. 'Can you unlock the door please?' Sally looks at me again, confused like she isn't the one in control of the car. I lean across her body to unlock it myself then step out.

I walk to the front of the car to inspect the damage. Sally remains open-mouthed at the steering wheel, staring out blankly.

'Fuck,' I say, looking at the car. The windscreen isn't as bad as I thought it would be, but there's a definite chip in the glass. The bumper is dented, the number plate split down the middle. It's not completely obvious from looking at the car that we've just run a man over in it, but it's bad enough that we'd have a hard time explaining the damage away.

I walk back around and open Sally's door.

'Get out,' I say and to my surprise she unclips her seatbelt. 'Actually, no. Park around the back where the car can't be seen from the road.' She nods as if this is the sort of thing she does all the time and drives us both around the back of the pub. We get out and walk into the pub in silence.

—

'Drink this,' I say, putting a whisky in front of her.

She shakes her head. 'I can't. I'm driving.'

I frown at her, as if she doesn't realise the damage that's already been done isn't going to be made worse by a little bit of whisky.

'I'll drive,' I tell her, pushing the whisky glass closer to her.

'You can't drink and dr—'

'I don't drink,' I say as I take a sip of my Coke.

Sally opens her mouth as if she's going to call me a liar but instead tips her head and shoots back the whisky in one.

We sit for a moment saying nothing. Things have become so much more complicated.

'You told me to drive,' Sally says. It's a statement but it hits like an accusation.

'Why was he there?' I ask.

She shakes her head. 'I don't know.'

'You didn't call him?' I ask, but Sally doesn't even reply to this, just keeps staring. 'Look,' I say, trying to sound like I know what the fuck to do. 'He's probably fine.'

'Alex,' Sally says blankly. 'We hit Alex.'

I tell her we did, and she starts shaking.

'Oh my god,' she says. 'I hit a man with my *car*. I *fled* from a *crime scene*. Oh my god. Oh my god.'

'Stop it,' I snap. 'People will hear. Just get your shit together.'

She keeps shaking but stops talking at least. The likelihood of anyone hearing us is pretty low; the pub is nearly empty apart from a few fat blokes sitting up at the bar watching the horse racing on the TV. They turned and looked us up and down as we walked in but quickly decided we weren't worth any bother and we haven't drawn their attention from the screen since.

This place reminds me of a pub I used to go to back in Birmingham. Shane liked it in there because they knew his old man; he'd died years before I was born but he was like a hero to Shane. Sounded like another piece of shit to me, but I'd never have told Shane that. Men like that love places like this; dark and desperate. No one to watch what you're doing.

This is the type of place where men can get away with anything. But we're not men. What if one of the blokes from the block is in here? I look around. It looked like a proper dive from the outside, the sort of place that if you walked in without a bloke, you'd usually turn back around and walk out again. But no one is watching us. I train my eyes back to the shit brown carpets and sticky tables while I let myself think. We might not have long before the police come looking for a wrecked car.

'Sally,' I say in the calmest voice I can muster. 'Did you tell Alex where I was staying?'

She shakes her head. 'Of course not.'

I scrunch up my nose. Did he follow Sally to me? It doesn't make any sense. Alex is a creep, but he's not exactly bothered about me. Not enough to trek into the woods to find me. Unless he was coming up to check out the building and we ran him over for nothing.

'What are we going to do?' Sally asks.

I shake my head. 'I'm thinking...' I rub at my eyes, desperately trying to find a way out of this.

'I have to tell someone,' Sally says, pulling out her phone. 'We need to call the police. Tell them it was an accident. They'll understand.'

I grab the phone from her hands and pocket it. 'Sally,' I say, so quietly she has to lean closer. 'They're not going to understand. You literally sent Alex flying over your windscreen then sped off – all while driving with a wanted criminal in your car. Do you not get it?'

'It was an accident. He came out of nowhere, you said it yourself—'

'No one is going to believe that. You know what he's been doing to me. They'll say you did it on purpose to punish him.'

Her eyes fill with tears as she shakes her head. 'They won't. We just need to explain.'

'You drove off!' I hiss. 'How the fuck can you explain that?'

She looks around the pub like someone might come to her rescue. 'I was in shock,' she says. 'People do all sorts of things when they're in shock. The police would understand.'

'The problem with you,' I tell her, 'is that you've no fucking idea what your lot can do when they want something.'

'My lot?'

I narrow my eyes. 'Yes, your mates at the police. You think they'll look after you because you're one of them but what will they say when they find out you've been helping me? That you ran over Alex for *me*?'

'I didn't do it for you!' she says too loudly. The bloke at the bar turns and glares at us and I smile, like nothing bad is happening. Sally watches and her cheeks redden.

'I don't know why I'm listening to you,' she says. 'This is insane. Every single thing I've done since I've tried to help you has got me into more trouble. I'm not ruining my whole life over someone like you—'

'Someone like me?'

'I'm sorry,' she says, her face darkening. 'But this is too much. I can't risk my whole life to save you from whatever it is you're running from. I don't owe you this—'

'So what, you're going to run back to your precious Nathan and tell him the truth?'

I hear my mistake too late and hope Sally's too panic-stricken to notice.

She screws up her face and studies me.

'How do you know Nathan's name?'

I shrug as if this is no big deal as a drop of sweat inches down my back. 'You told me in your emails.'

'No, I didn't.'

I shrug again, start messing with the beer mat on the table, ripping it into tiny pieces so I don't have to look into her eyes as I lie. 'Must have seen it on social media. I don't know.'

'I don't put anything about Nathan on social media.' Her voice is like ice.

I look up and meet her eye but say nothing. She searches my face as the horrible realisation she and I have more in common than she could ever have imagined dawns on her.

CHAPTER TWENTY-ONE

'You know Nathan,' she says. I frown. The truth isn't a simple yes or no, so I say nothing. 'How? How do you know him?'

I watch as I wait for her to piece it together herself, but she never will. There are so many moving parts to this story that I doubt anyone could piece it all together, not even me. Definitely not Sally.

'Natalie, for fuck's sake. What is it going to take for you to trust me?'

I sigh. 'Tell me something about Grace,' I say. 'That's a fair swap. You tell me where my baby is and I'll tell you how I know your fiancé.'

Sally looks at me open-mouthed like she's shocked I can be this transactional, but she truly has no idea what I'm capable of.

'I can't tell you that—'

'Then I can't tell you about Nathan,' I say, shrugging, like the information about Grace isn't a big deal at all, even though the mere mention of her name makes my heart physically ache, and it's taking everything I've got to stop myself from leaning across the table and beating the information out of Sally.

'Natalie,' Sally says, exasperated. 'I don't understand. Does this whole thing have something to do with Nathan?'

I wait a beat, then say it does, and that's enough to send her into a spiral. She looks around the pub again like there might be someone else here to help her and when she finds no one, she turns back to me. Her last hope.

Her face is pinched, like she's been forced into sucking a lemon as she tries, and fails, to work out what I mean.

'Come on,' I say. 'One thing. Just tell me one thing about Grace. You said they gave her a new name. What is it?'

Sally looks at me, weighing up her options here.

'Tell me her name and I'll tell you whatever you want to know about Nathan.'

The air stills around us as I watch Sally's lips moving.

'Abigail.' She says it so quietly it's barely a whisper but the name rings like a siren in my head. Abigail.

They stole my baby girl and named her Abigail.

I barely have a second to take in the information before Sally is talking again.

'Now your go,' she says, pulling her chair closer to me. 'Tell me how you know Nathan. Now.'

I take a deep breath. *Abigail.* Try to push her to the back of my mind. *Abigail.* She's a wholly different person now; it's not like I hadn't thought about it before, that my little girl had grown up with no connection to me, but hearing her name – a name I never chose for her, never even thought of – has shaken me to the core.

'How do you know Nathan?' Sally asks again. I shake my head and try to focus.

'I don't know him,' I say.

'But you did?' she asks and then, hesitantly, 'When?'

'Back then,' I say, waving to the past I try so hard every day to forget. 'What's she like?' I ask, unable to stop myself. 'Abigail. What's she like?'

Sally screws her face up in annoyance but I don't care about our deal any more. Hearing her name spoken out loud like this has opened up something in me. I need to know more.

'No,' Sally says, retreating back into her chair. 'We're not doing this. This isn't fair. She was taken out of your care for a reason—'

'She was stolen—' I spit, but Sally grabs me firmly on the arm and shushes me.

'Shut up,' she whispers. 'Now.'

I follow her gaze around the pub to the old, fat men at the bar staring at us over their pints. It feels like all eyes are on us and it hits me properly what a mess all of this is. I see Alex's crumpled body lying on the ground in the woods, I should care about him, really. It's not like I don't know him. I know him more than I ever wanted to. But still. Kind of think he got what he deserved. But I must admit it does put an added problem into the mix, and I can't afford to let things get worse than they already are. I need to be careful.

'We should go,' I say and before Sally has a chance to disagree with me, I'm storming out of the pub and back to the car.

—

'Natalie!' I hear Sally's angry voice calling across the car park to me.

She stomps towards me like an irate school mum; it's a shame she doesn't have children because she's the perfect type of woman to have them. Not like me. I think again of Grace – of Abigail. How much can I share with Sally before she realises the more I know, the more of a risk I am to her?

'We should probably stay on the move,' I say to her when she gets close enough. Her perfectly plucked eyebrows shoot up as if she doesn't understand. 'You know, 'cause we're on the run from the police?'

She steps back and shakes her head. 'We are not on the run.'

'We did run a man over, Sal,' I say with a laugh but then stop because she looks so completely broken by all of this. 'Nah, we're alright. Trust me, if anyone had seen us run him over, the police would have turned up by now.'

Her eyes flick over me like she's trying to work out if she can trust me. Truth is, I have no fucking idea if what I've said is true or not, but you'd think if someone had seen it and got Sally's number plate they might be here by now, wouldn't you? At least the fear of being caught has taken her attention away from the revelation that I know her darling fiancé, for now.

'When I went on the run last time,' I say, 'my face was plastered all over the papers within a few hours. Check online now, see if there's anything about Alex.'

We both get in the car and Sally takes out her phone. It's in a sensible thick black case and her lock-screen background is of a beach sunset that I can't work out if she's taken herself or if it's one of the stock ones that comes with the phone. For some reason, the idea depresses me. That Sally might have a stock photo as her background when she's got all the freedom in the world to go wherever she likes and take pictures herself.

'Nothing,' she says after a few seconds of scrolling. 'That's a good thing. Right?'

I shrug. 'You're the one in the police, not me.'

We hear it then, the scream of sirens.

We watch as an ambulance tears around the windy road, towards the woods.

'Look like someone called them,' I say, and Sally opens her mouth in disbelief. 'That's good news…'

She turns to me and frowns. 'Are you joking?' she says. 'How is any of this good news?'

I shrug. 'Well, the only person in those woods is Alex, so the only person who could have called an ambulance is Alex, which means…'

'He's not dead,' Sally says. She breathes out then closes her eyes.

Neither of us move while we take in this news. Alex isn't dead, we didn't kill him. That's a good thing, I suppose?

Eventually, Sally speaks. 'We need to go to the hospital.'

'What?' I yelp.

'Yes,' Sally says, warming up to her insane idea. 'The ambulance will take him to the hospital. We need to see if he's okay,' she says, then adds, 'and if he is, we need to make sure he doesn't tell anyone what we did.'

'Are you mad? We can't just rock up at the hospital asking after the bloke we smashed into with a car.'

'*We* can't, no,' Sally says, looking at me.

—

'This is the most fucking ridiculous plan,' I tell Sally for the thousandth time. We've left the car on the outskirts of Weston and got the bus into the centre to the hospital, in case Alex has already spilled his guts to the police and they're out looking for the car.

'It'll be fine,' Sally says breezily. 'Just go in and say your brother has been hit by a car and you want to see him.'

'What if they know he hasn't got a sister?'

'Then make something up,' Sally says. 'You're good at that.' She glances at me with a stony smile and I huff like a teenager.

'And when I come out, you promise you'll tell me more about Grace?' I say and she nods.

This is the agreement we've made. I go and work out what Alex has told the police and Sally tells me about my daughter.

We've already stopped at a corner shop to buy me a baseball cap, so if for any reason someone looks at the CCTV footage from the hospital later, I won't be recognised. I can't help but feel a little like I'm being used by Sally, but still – this is how the world works. I give her something, she gives me something back. She's been so wrapped up in the plan that she's not pushed me to talk any more about Nathan, which is good. It's giving me time to decide for myself how much I want to reveal.

'I'll wait here,' she says, pushing me gently on the back towards the hospital. She is sitting in the green bus stop on her own, looking as innocent as can be, while I head off like a lamb being led off to slaughter.

–

I walk up to the hospital with a buzzing in my head. Each step bringing me closer to Grace. Unless there are police on the door of Alex's room who recognise me and take me in straight away. This is such a stupid fucking idea. I hope he *is* dead. It's not like the maggot was doing anything good for this world. Or at least so badly injured that he can't open his stupid trap.

No. I don't. Because if he is dead – or almost dead – Sally could be done for murder. And I'd be her accomplice. For fuck's sake. Another crime to run away from. Sally's right that we can't just run away from this, though. If Alex has survived enough to talk, I need to remind him why it's not in his interests to do so. Plus, while I'm in here dealing with this drama, Sally's not bothering me about Nathan and if it leads to her giving me more information about Grace, it's a win–win.

'Hi,' I say, putting on my best fake posh accent as I approach the A&E desk. 'I think my, erm, brother has been brought in here. He was hit by a car. Can I see him?' I say, then add in my sweetest voice. 'Please?'

The woman behind the desk looks frazzled, her curly hair sticking out each side of her temples where it's escaped from her ponytail.

'Name?'

'Alex Harding,' I say.

She taps on her keyboard angrily and I wonder if she's not long finished a night shift and was supposed to go home but no one was there to cover her. Did they even bring Alex here? Maybe we did kill him. They don't bring dead bodies into hospital, do they? What would be the point?

'He's on the trauma unit,' she says and I breathe out. 'It's not visiting hours, mind, but go down and see if they'll let you in,' she says, handing me a map of the hospital and leaning around me to help the next person in line.

Well, sounds like we didn't kill Alex after all, doesn't it?

It's enough to push me forward as I follow the map around the snake-like corridors, the smell of antiseptic and old people permeating the air. I hate hospitals. I knew I wasn't going to have Grace in somewhere like this from the day I found out I was pregnant. People in places like this just can't stop meddling. They were always meddling with my mum and me when I was little and I didn't want that to be Grace's life from before she even came out, so I decided we'd manage things at home. I got a book off Amazon about pre-natal care and read bits, enough to know what I was doing, about vitamins and all that crap, and looked after myself while she was still inside me. I didn't need some jumped-up nurse telling a social worker I was knocked up.

I had Grace at home. It wasn't exactly how I'd have planned it but she came a few weeks early and didn't give me a lot of choice. It was perfect. Me and her on the floor of my bathroom. She just kind of slipped out, in the end. Everyone had told me it was going to hurt so much I would wish I was dead – and it did hurt, but… I don't know. I felt strong. Powerful. The two of us against the world.

If things had stayed like that, everything would have been fine.

But of course, things weren't fine. Because men like Shane exist and girls like me make stupid decisions that bring them to hospitals asking about the fate of men like Alex.

The swing door to the ward they've said Alex is on creaks as I walk through and I prepare to give my little speech about being his worried little sister again but when I walk through, the reception desk is empty and so I take my chance. The only people on the ward are a couple of old blokes in hospital gowns and a young woman sleeping. No police. There's one bay with the curtains drawn; I creep up to it, listening for voices that might be police or Alex's family but it's silent so I open the curtains.

There he is.

I let out a little laugh.

'What the fuck?' he says, lifting his bandaged head as his cheeks turn pink. 'What are *you* doing here?'

'Shh,' I say, pulling up a chair to the side of his bed. 'Are the police here?'

He shakes his head then winces at the pain. 'They'll be back, though.'

'What have you told them?' I ask.

'You mean, have I told them you hit me with a fucking car?' he snipes and I roll my eyes.

'I'm going to take that as a no, then.'

'Don't be so cocky,' he snaps. 'Like I said, they'll be back.'

I lean back in the chair and rest my feet up on his hospital bed. 'If you were going to tell on us, you'd have already done it.'

He glares at me but I know I've got him.

'Well, I'm going to have to tell them something,' he says. 'They want to know what I was doing in the woods. I got out of answering questions earlier because of this—' He points to his bandaged head. 'But that won't work forever. Want to tell me what the fuck is going on?'

I don't need to ask Alex why he hasn't told the police on us. If he was honest with them, he'd have to go into the details of our relationship and I'm guessing a scummy twat like him doesn't want to open his phone up to investigation. Not when it probably contains countless dirty pictures of girls like me with no other choice.

'We didn't mean to hit you,' I say, taking my feet off the bed and pulling my chair closer to him so I can whisper. I smell his musty breath, warm against my face, thinking of all the times he's breathed against me as he's taken what he wanted. Rank. 'What were you even doing there?'

'You told me to come,' he says, trying and failing to sit up.

'I didn't.'

He gestures towards his bag on the chair opposite and I pick it up.

'My phone,' he says, and I open the bag and rustle around. Remarkably, his phone has remained intact after the collision, and I hand it to him to unlock. He taps at it then holds the screen up to show me a WhatsApp message.

> Its Sally. Come to Bachmans Woods at 1p.m. today and I'll pay my debts ;) plus, Iv got a friend with me so double trouble XoXo

There's a picture of Sally attached – the real Sally. She's on holiday by the look of it, in a pink bikini. So, whoever sent this message knows I'm with her.

'This doesn't even sound like me,' I say but in truth, it kind of does. Not the real me, but the rank personality I put on for Alex to get my own way. 'And it's not my number.'

'You're saying you didn't send me this?' he asks. I take out my own phone and write down the number to show Sally later then delete the WhatsApp from Alex's phone so there's no evidence for the police.

'Of course I didn't, you fucking idiot.'

He winces as he glares at me. 'Something got sent to the office for you,' he says, nodding towards his bag on the floor. I open it up and pull out a brown envelope with my name on it. No stamp, no address.

'Did you see who delivered this?'

He shakes his head and again winces at the pain.

'Stop doing that,' I tell him. 'It obviously hurts.'

He narrows his eyes. 'Like you give a shit.'

I open the envelope and quickly read the typed note.

ONE LAST CHANCE – THE MONEY WILL COME BACK IF YOU KEEP HIDDEN. DON'T BE STUPID, "SALLY JONES".

I take in a sharp breath and close my eyes. This is such a mess.

'What the fuck is going on?' Alex hisses. I open my eyes to see a fleck of spit jumping from his mouth and landing on his weak chin. 'You and that other Sally Jones bitch – did you plan this?'

I shake my head. 'No, you jumped out of the car like an idiot and got yourself hit. Trust me, I don't want the police sniffing around me any more than you do.'

We both stop as we hear the crackle of a police radio.

'Look,' I say. 'I'm sorry you got hurt. But you're fine, really, aren't you?'

Alex glares at me as I fold the note back into its envelope and peer around the curtain to check for police.

'You can't say this had anything to do with me. Say you were coming to check on the building work's progress and you didn't see the other car.'

Alex scoffs like this story is totally unbelievable but I don't care.

'Say something, anything, but leave us out of it and I'll make sure no one ever finds out about your side-business, okay?'

He sighs but nods, closing his eyes as he winces in pain. For safe measure, I pocket his phone then creep out from behind the curtain before he notices I've taken it.

As I do, my phone buzzes in my hand and a familiar number appears on the screen.

> I'm glad you got my message. In case it wasn't clear – I want you gone. For good, this time. If not, you'll find yourself in a hospital bed too – only unlike your mate, I'll make sure you never make it out again.

I close my eyes, as if darkness might stop the threat from being real. I look around me as if I'll see the person behind the message watching me without remorse, but the hospital room is still only occupied by old half-dead people. The person behind this message isn't here. But they know where I am; they know what I'm doing. How did they get this number? I had chucked out my old phone once I left the flat and got this old brick from the guy I buy all my dodgy gear from. No matter what I do, I'll never escape. But there's no time to think too deeply about any of that because two police officers are a few feet away from me now, and if I can't make it past them, I may as well be dead.

CHAPTER TWENTY-TWO

I pull my cap down and creep behind the officers, imagining they'll whip around any minute and grab me, but they don't turn around. The female officer laughs merrily at a joke the receptionist makes while the bloke talks on his phone.

'We're about to interview him, boss,' he says, then pauses. 'No, he was concussed.'

I pick up the pace and keep my head down, walking so close behind them that I can smell the apple shampoo of the female officer's hair and then I'm past them, breathing a sigh of relief.

'Excuse me?'

It's the male pig, calling from behind me. Shit. I keep walking, pretending I can't hear him.

'Hey,' he calls, then I feel an arm on my shoulder.

This is it. Things are finally catching up with me.

I spin around to face him – to face everything I've done.

'I think you dropped this,' he says, a small smile on his face. He holds up the envelope that must have fallen out of the side of my bag.

'Oh,' I say, waiting for him to say something else – to notice who I am. But he doesn't, just smiles at me then frowns as I fail to take the envelope from him.

'Thank you,' I say, shoving it into my back pocket like it's nothing.

His radio beeps and he turns away to answer it, letting me scuttle away. Free.

I snake my way back through the hospital corridors with my head down, the cap pulled low to cover my face. The woods were a set-up. They knew Sally had come to see me so they sent Alex to warn me off with that stupid photo of her and the nasty note. They want me to know that they're watching. They're always watching.

One thing is for sure – I'm no longer safe. If they found me at that hideout, they must have been tracking Sally. She's my only connection with the outside world and she's also the only person stupid enough to let herself be tracked like this without having a clue.

I've got no choice. I'm going to have to get rid of her. I'm not safe with her around.

But she's my only connection with Grace.

How can I give that up? I've not let myself dream of getting to know my little girl again and yet here is Sally with the information that might allow me, one day, to find her. To be in her life. I have to find out more before I disappear again.

As I walk out of the hospital, the fresh air hits me and clears my mind. Sally doesn't want to be caught up in all of this. She wants a nice easy life; but she does want to know about Nathan. Maybe that's enough to use as a bartering chip. I scan the bus stop across the road from the hospital and try to get Sally's attention. I don't want to be out in the open with her, not again. I wave my hand and eventually she spots me. I signal to go around the side of the hospital, and wait there for a moment until she appears.

'Well?' she asks, her eyes darting around like we might be rumbled at any moment.

'One thing about Nathan, one thing about Grace,' I say, and she furrows her brows in confusion.

'What?'

'My go. I knew Nathan ages ago and I don't know him now,' I say and she steps back.

'What are you talking about? What happened in there? Did you see Alex? Is he okay?'

I shake my head in frustration. 'He's fine. Bashed up a bit but he's fine.'

Sally breathes out and holds her hand to her heart like she's trying to stop the pounding.

'Did you talk to him? Has he spoken to the police?'

I shake my head, impatient. 'He won't tell the police.'

'How can you be so sure?'

'Fuck's sake, just believe me!' I shout and Sally takes a step back, hurt. 'I'm sorry,' I say. 'But I don't have time for this. You don't need to worry about Alex, okay? He isn't going to say shit, so you can go back to your perfect little life and forget any of this ever happened.'

Sally shakes her head. 'What are you talking about? I can't just walk away, Natalie. You're clearly still terrified of something – or someone – and you've told me you know my fiancé but not how or why—'

I grab her shoulders in frustration.

'We don't have time for this,' I snap. 'You want to know about Nathan, I want to know about Grace. So tell me something.'

She purses her lips and folds her arms.

'Tell me something now or I'll phone the police myself and tell them exactly who hit Alex.'

She opens her mouth, hurt and then betrayal hardening her features.

'I mean it. You want to get out of this without the police coming after you, tell me something about Grace. Now.'

Sally's cheeks redden as understanding blotches her pretty face. She knows what type of person I am now; she sees me for who I really am and she doesn't like it one bit.

'Grace…' she starts, then stops as if doubting herself so I wiggle the phone in my hand, a veiled threat. 'She was adopted by a family I know.'

I squirrel the information away for later and give Sally something in return. 'Nathan worked for Shane. He promised he'd help me leave him,' I say, and when the words are out there in the world they hurt all over again.

Sally takes another step back but I pull her in closer so no one can see her from the side of the building.

'Nathan…' she says, bewildered, like she can't comprehend a world where her beloved fiancé would ever be involved with someone like me. 'This doesn't make sense.'

'One thing about Grace,' I say, sternly.

Sally closes her eyes and rubs at her face. 'She lives in Warwickshire.'

My heart bashes against my ribs; now we're getting somewhere. 'Does she know who I am?'

Sally shakes her head then twitches her lip. 'No,' she says and though I suspected it, the fact hurts. 'She knows she's adopted. She's too young to know… you know, everything else.'

'Who adopted her?'

'No,' Sally says, some of the fight back in her eyes. 'One thing about Grace, one thing about Nathan. That's the deal.'

Something about the look she gives me, like she's made a deal with the devil himself makes me say it. I want to hurt her in that moment, to make her understand.

'I loved him,' I say. 'I really fucking loved him.'

She laughs, loud and shrill, and it makes me want to slap her stupid face.

'You *loved* Nathan?' she says. 'This is… This is fucking ridiculous. You're making this up. Nathan would never work with someone like Shane. He'd never be with someone like *you*. You're just saying all of this to make me—'

'To make you what?' I ask. 'What could I possibly have to gain from any of this?'

She shakes her head and looks at me like something on her shoe. 'I don't believe you.'

'I don't give a shit what you believe,' I say but I do. Of all the reactions I thought Sally would have when I told her the truth, this wasn't one of them.

Sally shakes her head and takes out her phone, unlocking it.

That's when I see the picture on her background.

It's him.

My god. It's him.

And just like that, things start clicking into place.

Sally isn't a safe person for me, how could I ever have thought otherwise? This little naïve nice girl act – that's all it is. An act. She knows exactly who Nathan is – she's been in on everything the whole time. How could I have been so stupid?

'I'm calling Nathan now,' she says and for a moment I can't even react. 'We need to sort this out. Whatever is going on, you need to stop lying.'

'Great, I'll call the police while you're at it. Better yet, why don't I just go back into the hospital and get them down here?'

She pauses, stunned by the venom in my words, and I take the opportunity to grab the phone from her hand. It crashes between the two of us on the floor and I kick it, hard, sending it soaring around the corner. Sally chases after it, her long hair flying in the wind as she disappears from view, and I take my chance to do what I do best: I run.

PART THREE

CHAPTER TWENTY-THREE

SALLY

I grab my phone from the floor and spin back around, expecting Natalie to be right behind me but she's not. I take a second to breathe, to try and get back to some sort of sanity before walking around the corner.

She's gone.

Natalie is gone.

I shout her name, scanning the side of the hospital for any sign of her, but there's nothing. It's like she was never here at all. The phone beeps in my hands and I open WhatsApp. It's her.

> Don't look for me. Go back to your cosy little life and forget any of this ever happened. It'll be better for both of us this way.

I stare at the message, then delete it before I can decide otherwise. I should get rid of this phone. What if the police are tracking it right now? I put it on the floor and go to stamp on it – but stop. Destroying evidence, that would be yet another crime. They're racking up. The thought makes my face hot and I know I need to get out of here. Now.

—

I've walked back into the centre of Weston, every step feels risky as I listen out for the sound of sirens, thinking of Natalie's threat. I've turned the Wi-Fi off my own phone, thinking ahead to a trial where they could use my geo-location as evidence against me. How the fuck is this my life now? How am I thinking about being on trial? When I reach the beach, I let myself slow down, the sea breeze catching my hair and cooling down my sweaty face.

Natalie said Alex was fine, but can I trust her? When she came out of the hospital, something in her had changed. Perhaps she knew she was an accomplice to murder. The thought makes me feel sick and I double over on the beach and take in big, heaving breaths of sea air.

Fuck fuck fuck. I should have gone in myself. I should have seen Alex with my own eyes. Why did I ever trust her? The look on her face when she threatened to call the police on me was one of pure contempt. I was an idiot to think I could ever help someone like her. She's a monster. I force myself to remember what she did to Shane Blackwell – the violence of his murder – the way she abandoned Abigail as a tiny baby and never came back for her. She said it wasn't her choice, but do I believe her? After what I just saw in her, I don't think I do. People make their own choices, and she made hers. Just like my mum did. I've spent a lifetime knowing women can be crueller than anyone ever admits and yet I let myself get reeled into Natalie's story where she wasn't the villain but the victim.

I sit on a rusting iron bench on the seafront, running my fingers over the engraving.

> For James, my darling husband who sat by
> my side here for forty years.

It brings hot tears to my eyes. Such a simple sentiment and yet one I don't know if I'll ever be able to say about Nathan. How can I go home to him now? Do I even want to? I told Sally I didn't believe her that she knew Nathan – *loved* Nathan – but now that I'm alone, I can't be so sure.

I think back to the time Sally was talking about. She said Nathan was working with Shane. He was a West Midlands officer back then, with our neighbouring force. If Natalie had murdered Shane five miles further west, it would have been on their patch, not ours. It was before Nathan and I got together and all I remember is that he was away a lot that year. He was on some sort of detective training or something. It's hard to remember the specifics. We didn't get together until six months or so after Shane Blackwell's case.

I try to think what he's said about the case over the years. We've talked about it often because it had such an impact on me. Taking that call from Natalie and believing her when she told me she was the victim. How it changed my view of my job, my view of the world. She became another person who had lied to me and let me down, another woman who didn't give a shit about the life she'd created. I knew I was displacing my anger over my own mother abandoning me and projecting it on to Natalie, but the nation shared in my disgust and so I felt justified. Nathan shared my rage; he saw what being abandoned by my mum did to me when I was little – he knew what it would do to Abigail as she grew up.

Has Nath ever given me any reason to believe he knew more about Natalie's case than he should? No, there's nothing. Whenever we've talked about it, it's been about my point of view. Me taking Sally's call, the sound of Abigail crying in the background. Me believing her to be a good mum who wanted the best for her little girl. Nath has never given me any reason whatsoever to think he knew the woman on the end of the phone. None at all.

Maybe Natalie is lying. She could have found Nath's name out from somewhere — anywhere — and used the information against me. She wanted information about Abigail, that's all. That must be why she targeted me in the first place, to get close to her little girl. But how could Natalie have known I have a relationship with Abigail? It's not like we're related — the fact Amanda became so close to Chloe and Matt is an anomaly — not the usual process at all.

Unless Natalie has followed me since the night I answered her 999 call. She could have seen something on social media. Chloe, Abigail's mum, does post things on Facebook occasionally, pictures of us sitting in Amanda's garden. Nathan always stays out of the photos — like a lot of officers in his position he avoids social media like the plague — and Abigail's face is always covered with a big yellow smiley face, but if Natalie was watching us all closely, she could have put it together that the five-year-old girl I'm often with is the daughter she abandoned.

I wring my hands and look out to sea as the waves crash angrily in the distance.

I don't know what to do.

I always know what to do. It's one of the things I'm known for at work; I'm unflappable, that's what they say.

Good in a crisis. And yet right now, the biggest crisis I've ever been in sees me totally paralysed.

I wish my dad was here. Though would I be able to be honest with him about the mess I've got myself into? He'd be so disappointed in me – something he never was. Well, not until I started seeing Nathan, anyway.

It never made sense to me that he didn't like us getting together. He'd known Nath since he was a baby; there had always been an ongoing family joke that one day we'd get married, right from when we were toddlers. Dad and Amanda would laugh about it as we held hands in the garden. But Nathan was never interested in me like that; I knew he saw me as his little sister.

Until that summer when everything changed.

The summer right after Shane Blackwell's death.

I turn the fact around in my mind, trying to connect the dots.

But I can't. My brain isn't working – I can't think straight. I can't ask my dad about any of this, obviously, and I can't go to Nathan. How could I trust what he tells me? The only other person I could speak to is Amanda.

As soon as I have the thought, I feel better. Amanda will know what to do. I can't tell her about Natalie, or the accident, or any of what she said about Nathan. But I can ask her what was going on with my dad back then, why he was so against Nathan. If it has something to do with Natalie, she'd know.

I stand up, pleased I have some part of a plan and start to walk back to the pub where we left my car.

—

By the time I reach the pub, my earlier pragmatism is gone. I look at the wrecked front bumper, sighing. I can hardly drive it home like this. The best thing that could happen would be for it to get nicked. Then, if it ever does come up that it was involved in Alex's accident, there is reasonable doubt that it was being driven by me at the time. Cars get nicked all the time; if I say it broke down and I left it to walk into town, no one will doubt me.

God, am I actually going to do this?

I drive back towards the town, keeping an eye on my mirror all the time as if the police might appear behind me and when I'm near enough to walk the rest of the way, I pull into a side road I saw earlier. I check for a camera: nothing. The area is run down and craggy as anything, but it's got a big steep hill that ends in a cavern.

I think about all the criminals I've read about over the years — always believing myself to be nothing like them. Did they start off like me? One minute you're a normal person, making good decisions — then, bam — you're thinking about rolling your car off a bloody cliff to cover up a hit and run.

This is madness. But what other choice do I have? I've come too far to do the 'right' thing now.

I get out of the car, grab a heavy rock, place it on the accelerator then watch as my car, my lovely little banged-up car I've had since I was twenty-five years old, crashes headfirst into the abyss. Wiping away all evidence of my previous crime.

—

'Amanda?'

'Hi, love.' Amanda's breezy voice comes down the phone as I walk through the town centre. It's early evening and though the high street wasn't exactly a beacon of respectability earlier in the day, it's a real dive now. Men with bloated overhanging bellies criss-cross the street in front of me, fags hanging out of their mouths as they shout to each other with warm, beer-tainted breath. I pull my denim jacket tighter around me as they laugh.

'Where are you?' Amanda asks.

'Weston,' I say and I hear her sigh, heavily. 'I need a favour.'

'Of course. What do you need?'

'My car's broken down. I'm a bit stuck. I can't call Nath, he's on shift. Is there any chance you could come and get me?'

For a few moments Amanda doesn't say anything, and I pull the phone from my ear to check she's still connected.

'Yes, no problem. I'll leave in a sec. What happened to the car?' she asks.

'I don't know,' I say, sheepishly. 'It was a couple of hours ago, I was parked up by the old pier to walk along the beach and it just conked out. But look... it's been stolen—'

'Your car has been stolen?'

'I'll explain when you get here, okay?' I say, my cheeks flaming because why didn't I just say it was stolen in the first place? I'm such a terrible liar.

'Have you reported it?'

'Not yet,' I say.

She pauses, sighing. 'Well, call it in. You'll need it for insurance,' she tells me, and I tell her I will, as if claiming for the old rust bucket on insurance is what I care about right now.

'I'll be with you in an hour and a half. Try not to get yourself into any more trouble before then, okay?' She laughs warmly, and we hang up.

After she's gone, I look up and down the high street for somewhere to wait. The wind from the seafront is blowing through the buildings and suddenly I'm freezing. All around me, women in not enough clothes and drunken men zigzag in and out of the grotty-looking pubs and bars. I get to McDonald's, where a group of teenage boys on bikes gather around the front ominously. I keep my head down but then I hear my name being shouted.

'Oi! Sally!'

I look up.

'Tiff,' I say, relief washing through me at the sight of a familiar face.

She cocks her head to the side as she studies me.

'Who's your mate?' the boy stood closest to her asks, slinging his long skinny arm around her shoulders. Tiff shrugs him off and walks towards me, to the chorus of 'stuck up bitch' shouted by the spurned boy.

'What are you doing here?' she asks as she ushers me away from the group.

I want to talk to her, but the lads are still watching us and I feel trapped, like we're in a goldfish bowl. Tiff must sense this and pulls me into an overly hot red-fronted cafe across the road. We sit at a white Formica table that's sticky with old ketchup, and I resist the urge to ask a waitress to wipe it clean.

'You look a bit rough,' Tiff says, arching her brushed up eyebrows at me and, despite everything, her directness makes me smile.

'It's not been the best day,' I say, picking up the laminated menu from the table. Also sticky.

'Never is round here. You gonna buy me something?' she asks, picking up the menu.

'Get whatever you like.'

Her face lights up greedily as she scans the options and beckons a young waitress over.

'What can I get you?' the girl asks.

'Cheesy chips – but do 'em properly, like. None of that cold grated cheese on top. I want it melted,' Tiff says with a grin. The waitress smiles like this is what she's used to. 'And a Coke. And a hot chocolate,' Tiff says before looking at me like a little kid waiting for permission, then adds, 'and a chocolate ice cream sundae after.'

I say thank you. The waitress probably thinks Tiff is my child, or a niece who I don't see often and like to indulge when I do. If Tiff were my niece, I'd take her somewhere nicer than this, but she's happy enough.

'Black coffee for me, thanks,' I add to the order, the idea of fatty food turning my stomach, and the waitress disappears again.

'Don't be thinking you'll be having any of my chips if you ain't ordering nothing yourself,' Tiff says as I check my phone for updates on Amanda's journey. She's still got over an hour to go and I'm glad of Tiff's company in the meantime.

'Don't worry, your chips are safe,' I tell her as I lean back into the cheap chair.

'So, what are you doing here?' she asks. 'Not looking for your mate still?'

I shake my head, unsure of how much to reveal to Tiff, but in the end I think what's the harm? She's a teenager, she doesn't care about any of this. She'll likely forget it the moment she walks out of the door and back to her straggly mates.

'I found her,' I say, and Tiff's eyes go cartoon-wide. 'But it turns out she's not—'

I stop as the waitress appears by the side of the table with a plate piled high with steaming hot cheesy chips.

'Yes!' Tiff says. 'That's what I'm talking about.' She pulls a chip from the middle of the pile, yellow stringy cheese oozing from it so hot she squeals as it burns her fingers. She puts the chip back down impatiently. 'Turns out she's not what?'

'Who she said she was,' I say, simply.

'So where is she now?'

'Gone,' I say.

Tiff frowns before going back in for a chip, blowing on it as she thinks. 'So why you still here?'

'I'm waiting for a lift home.'

'Where's your car?'

I look away from her, my eyes trained on the old couple a few tables away from us as they smile at each other over their egg and chips. The smell of the deep fat fryer wafts under my nose and suddenly I'm ravenous. I lean across and nick a chip from Tiff's plate before she can slap my hand away.

'Oi!' she yelps, looking genuinely angry but at least I've distracted her from asking about my car. Why can I lie to Amanda but not to this teenager I barely know? There's something about Tiff that demands honesty. And I know if I start to tell her the truth about the car, it'll all come tumbling out.

'I've got to go,' I say suddenly, getting up. 'I'll get this, enjoy the food. I'll see you around, okay?'

Tiff opens her mouth then shuts it, as if I've upset her but she gives me a little wave, and shoves another chip into her mouth.

CHAPTER TWENTY-FOUR

I'm freezing by the time Amanda arrives. I waited for her on the beach, needing to be on my own, away from Tiff's inquisitive glare, but regretted my decision once I was sat on the cold, wet sand with the wind whipping my face and a never-ending parade of evening dog walkers passing me by. By the time Amanda's car appears, I'm too bloody cold and fed up to feel anything but pure relief. It's not until I'm strapped into my seat that I realise how difficult this conversation is going to be.

'Alright?' Amanda asks, concern written across her face as I slip into the passenger's seat beside her.

'Yeah,' I say. 'Thanks for coming to get me, I'm so sorry you had to drive all this way.'

Amanda shakes her head and tells me it's not a problem and we sit in silence for a few moments as she navigates the one-way system out of Weston town centre. I focus on the people we pass as we go; hen- and stag-dos spilling from the pubs, chanting silly songs with veils over their faces for the girls and over-filled pints dribbling down the arms of the men. They're here to have a good time, nothing more complicated than that. I close my eyes and imagine my life is so simple. I haven't had a hen-do – I couldn't imagine anything worse, or at least that's what I told Nathan when he suggested it. The truth is I was worried the friends from work would feel obliged to show up and spend the night wishing they were elsewhere. The only people I could have counted on wanting to celebrate would be Amanda, and Chloe.

At the thought of Chloe, I wince. One of the only people in the entire world I can call something close to a friend – and look how quickly I betrayed her.

'What's up?' Amanda asks.

'Nothing,' I say shifting awkwardly.

What would she say if I admitted I'd told Natalie Pierce all about Abigail? I even said where the little girl lives now – though I don't think I said anything more specific than 'Warwickshire' as her location. God, I hope not. But I would have said anything to Natalie to get her to tell me more about Nathan. Anything. Shows how good a person I am when it comes down to it, doesn't it? Willing to sacrifice the safety of a child for my own needs.

'So, are you going to tell me what's going on or are we going to spend the next hour and a half in silence?' Amanda says as we pull onto the M5. 'Nathan called earlier. I didn't tell him about this little trip.'

'Thank you,' I mumble, like a naughty school child.

'And I won't tell him either, if that's what you want. You know you can trust me, Sal. I know he's my son, but you're as good as my daughter. You always have been.'

Amanda's not a touchy-feely type of person so this statement hits me hard; so hard I find it impossible to keep my tears from falling, hot against my cold cheeks.

'I promised your dad I'd look after you,' she carries on as she switches into the slow lane behind a huge lorry. 'And I can't do that if I don't know what's going on. So, please, talk to me.'

I nod, blurry-eyed at the mention of Dad.

'Why didn't Dad want me to marry Nathan?' I ask and she glances over at me for a second, frowning.

'Of course he wanted you and Nathan to be together—'

I shake my head. 'He didn't. He told me when he was in the hospital. I didn't tell you and I'll never tell Nath, but it was one of the last things he said to me.'

Amanda frowns but doesn't say anything for a moment, as if she's digesting this new information.

'What exactly did he say?' she asks.

I sigh and close my eyes, not wanting to look at Amanda when I say the words.

'He said I shouldn't marry Nathan. That I should find someone else, anyone else. But not him.'

I open my eyes and flick a glance over at Amanda. A red rash crawls up her neck, and I wonder what she's thinking. I remember how I felt when I heard those words from my dad, the man who had watched Nathan grow up; who cheered in the audience at his police graduation ceremony; who joked for years and years with Amanda that one day he'd be walking down the aisle proudly giving me away to Nath. And then, this.

'He wasn't well, love,' Amanda says after a while. It's the same explanation I've given myself.

'I know, but he seemed so lucid when he said it.'

She sighs. 'Dementia does funny things to the brain, you know that. At the end, your dad was saying all sorts. I'm sorry you had to hear that from him, but I don't think you should pay any attention to it. He wouldn't want you to. He loved Nathan, he always wanted the two of you to get together. You know that.'

She's telling me the exact same thing I've told myself over the years, because she's right. Dad did love Nathan, he loved how the four of us – me, Nathan, him and Amanda – formed our own little family when my mum abandoned us, and when the two of us finally got together as adults, I thought he would be thrilled. But he wasn't. He never said anything directly to me about it – not until that day at the hospital – but I knew how he felt. I could tell by the way he'd leave the room whenever Nath and I would walk in, hand in hand, with the excuse that he needed to do something in the garden – or make a cup of tea. It was like he needed a few moments to rearrange his face before he could see us together. I ignored it – never mentioned it to Nath or to Amanda, because I didn't want it to be true. It was a taint on this new, perfect relationship I'd dreamed about having for my whole life and I couldn't quite bring myself to admit that my dad felt that way.

'But I don't understand why he would say it,' I say now to Amanda. 'Yes, the dementia did make him come out with some strange things at the end. But they were usually mix-ups – not completely new and strange opinions.'

'I don't know, love. Did you not ask him at the time?' She looks at me and I shake my head.

'A nurse came in and interrupted. I planned to talk to him about it the next day, but that was when he...' I stop, the memory of the day Dad died still too unbearably painful to talk about.

He'd been diagnosed with dementia years before but he'd seemed so well with it, it almost faded into the background of our everyday lives. But then things went downhill rapidly; he went into hospital and not long after we got that awful phone call. A heart-attack in the middle of the night. He never woke back up.

Amanda nods, understanding what I can't put into words. I was too late. Dad was gone. It still feels unreal, even six months on, that the man who was my whole world no longer exists. He's not just in another room, making a cup of tea, watching telly – he doesn't exist at all. It was bad enough after my mum left, but I've always known she's probably still out there somewhere, having a lovely life without me. But with my dad, it's different. He didn't choose to leave me; he never would have.

'I wouldn't think about it too much, Sal. What's brought this back up for you, anyway? Are you having cold feet about the wedding?'

'No,' I say sharply. 'Not exactly. It's just... I don't know. It's made me wonder whether perhaps Dad knew something about Nathan. Something bad...'

Amanda switches off the radio, which until now has been playing quiet, melodic pop music in the background, and says nothing for a moment, just lets us drive in the silence. Around us, cars whizz by. I check Amanda's speed gauge, a steady seventy miles per hour.

'Nathan loves you,' she says.

'I know he does.'

Amanda nods, like my agreement means the matter is closed.

'You know, when he was away doing training at West Mids...' I start to say and Amanda sighs. She cuts across into the fast lane and I watch the speedometer creep to seventy-five, then eighty, then eight-five. 'What was he doing?'

'What do you mean?' She flicks a glance in her wing mirror before undertaking the car in front of us.

'What sort of training? He was already a detective, so it wasn't his detective training. He was away a lot, wasn't he? I don't know what sort of training that would be.'

'I don't remember,' she says, not looking at me.

'You must, it was the same year as the Natalie Pierce case,' I add, watching for a reaction, but Amanda's attention is firmly on the road. 'Nathan was still at West Mids and he was busy – a lot – always saying he was "away" with work, but he never explained what he was doing. You must remember,' I say again.

We speed along the motorway as I wait for Amanda's reply but she stares resolutely ahead as the speedometer climbs: eighty, ninety, ninety-five miles per hour. I hold onto the door handle, suddenly anxious we're going far too fast. I'm about to say something when the flashing fifty signs come into view and the motorway is lit up with the red lights of the slowing cars ahead of us as Amanda slams on her brakes.

'Shit,' she says under her breath as we finally come to a stop a couple of centimetres shy of smashing into the back of the stationary traffic.

'Are you okay?' I ask. Amanda's cheeks are bright pink; she looks ruffled in a way I've never seen before. She's usually such a calm driver.

'I'm fine,' she says, shaking her head with a smile like this is all some silly joke between us. 'Got a bit carried away. It's a long way to come back and forth on the same motorway, you know.'

I burrow down into my seat, guilty I've dragged her out to pick me up like this but not guilty enough to not press her for information. 'So, Nathan – that training course. What was it for?'

'I've no idea,' she says with a smile. 'You'd have to ask him, though I doubt he'd remember something from so long ago.' She waves her hand as if whatever Nath was doing back then is trivial. 'Now, are you going to tell me what this is really about? Why you're suddenly so obsessed with Weston? Why I've driven an almost 200-mile round trip to fetch you?' She turns to me, eyes hard and the blush from her cheeks long gone. 'Because to be honest, Sal, I'm more than a little worried about you, and all this questioning about Nathan, it doesn't sit right with me.'

I sit lower in my seat as the traffic in front of us remains at a standstill. I'd forgotten what it feels like to be the subject of one of Amanda's interrogations and I don't like it at all. Growing up, Nath and I said we had the absolute worst of it with his mum and my dad both being detectives. We couldn't get away with anything. We played knock-door-run once and got caught by the neighbours. When they came to Amanda's to tell on us during a barbecue with everyone in the back garden, we were pulled away from the fun and sat down in Amanda's study for questioning by them both. Though Nath came up with a convincing cover-up story on the spot, I was useless and confessed every last detail in under sixty seconds. It became a family joke that if I ever committed a crime, I'd probably walk into the nearest station and turn myself in before the police even had it on their radar. But I'm not laughing now.

'I found my friend,' I say before I can stop myself. 'You know, the one with the same name as me.'

'The one I told you to stay well clear of,' Amanda says, her eyes not leaving mine.

I say nothing.

'And?'

I sigh. 'She wasn't who she said she was.'

Amanda doesn't say 'I told you so' but I can feel the statement between us. 'So, who was she?'

I shake my head. 'It doesn't matter. Just not... not the person I believed her to be. It got me thinking,' I say, trying my best to stick to my half-truth. 'About how much we ever really know about anyone and that made me remember what Dad said about Nathan and how strange that time was when Nath was never around and then he transferred straight out of West Mids to Warwickshire, even though he always swore he'd never serve in the same force as you—'

The words come out harsher than I intended, and Amanda looks stung.

'Not like that,' I say, trying to claw the hurt back. 'But you know he wanted to stand on his own two feet. He always said he'd get so much shit for serving in his "mum's" force. You know what it's like.'

Amanda sighs, and looks away as the red brake lights from the car in front of us light up her face.

'So, why did he transfer?' I ask.

I've asked Nath this before, of course, but never with much interest. It all happened around the same time he and I got together, so everything else felt so unimportant in comparison. But was it all connected to Natalie and Shane? Did Nathan get caught up with them and have to move forces? But that doesn't make sense because they mostly operated in Warwickshire – so why would he transfer into that area? I suppose the obvious thing is that Shane was dead by then, so any threat he held over Nathan would be gone. But still, it doesn't add up.

'What did Nathan tell you?' Amanda asks and I know then that I'm onto something.

'Nothing,' I say. 'That's why I'm asking you.'

She sighs. 'I shouldn't be telling you this,' she says. 'But if you really must know, Nathan was working undercover—'

'What?'

'Let me finish. He hated it. Didn't suit him at all. He wanted a clean break after that, it bruised his ego, to be frank. So yes, as much as he always hated the idea of being in "mummy's force" I think actually, after that experience, he wanted something easier.'

I don't say anything, my mind turning over this new information and trying to slot it into place.

So maybe Nathan did know Natalie. But he wasn't working for Shane like Natalie believed.

He was undercover.

But the real question is, was it truly a bruised ego that forced him to give up undercover work, or was it Natalie?

Jesus Christ.

Nathan and Natalie. That's what this is all about.

I loved him. I hear Natalie's words in my head.

She loved him.

The question is, did he love her too?

CHAPTER TWENTY-FIVE

It's almost dark by the time we arrive on my street. Amanda and I haven't talked for the last hour of the journey. Instead she put on Radio Two and we listened to Sara Cox chat away frivolously with guests from up and down the country about how they were spending their Thursday nights. I tried to imagine being one of them, standing at the cooker making a spaghetti bolognaise with a bottle of red wine on the go, not a care in the world. But instead, I'm stuck in a car with my mother-in-law after finding out her son has kept something huge from me.

'Sal,' Amanda says, pulling me out of my self-pity spiral. Her hand reaches for me across the car and she gives my arm a little squeeze. 'I know you're probably annoyed Nath didn't tell you about this undercover business before, but you know that he couldn't. I didn't even know he was undercover until he came off the job. You know that's the rule with that sort of work.' She smiles at me weakly and I nod.

She's right in one way; Nath wouldn't have been able to tell me about his involvement with Natalie Pierce at the time, but in the years since, why wouldn't he ever mention he worked the case? Unless he had something to hide.

'Yeah,' I say to Amanda as I grab my bag from the footwell. 'I know. It's just... A lot. I thought I knew everything there was to know about Nath.'

Amanda nods, her smile fading a little. 'You do, love. Everything that's worth knowing, anyway.'

'Thank you for picking me up,' I say, wanting to be away from this conversation now. To have some time to think, alone.

'It's no problem. Before you go,' she says as I pull open the car door and then gestures for me to close it again. 'I've got some good news.' Her smile brightens as I wait to hear what good news she could possibly have. 'I know how fed up you've been since you've been put on "special leave",' she makes air quotes with her fingers and frowns, mirroring my exact feelings about the 'leave' I've been given. 'So I had a word with Holden and got you put back on your normal job. You're expected back on the day shift on Monday.' She smiles widely.

'Oh wow,' I say. 'Thank you. I... I thought they wanted me off the operational side for the foreseeable.'

'No,' Amanda says. 'Well, if they did, they don't now. I sorted it, so you don't have to worry. So, put all this Weston stuff – and Nathan business – out of your head and get back to normal. Okay? The wedding is in just over a week's time! You should be focusing on that.'

She takes a breath, and leans over to open my door. 'Now, give Nath my love, and Sal, I really wouldn't bring up the undercover stuff with him, okay? It wasn't a good time for him.' She sits back, leaving my door open, letting in the cold night's air. 'He'd hate you to know how badly he failed as an undercover cop.' Amanda laughs softly, as if the whole episode is nothing more than an embarrassing story for her son. 'You know he likes to play the big, brave bloke around you. Always has.'

I tell her I won't say a word and we say our goodbyes. As I watch Amanda's car drive up the street, I try to tell myself that maybe she's right. Maybe the only reason Nathan misled me about his past is because it's embarrassing. Maybe that's all this is.

—

Of course, that level of self-deception can only last so long and by the time Nathan is home a couple of hours later, I'm basically climbing the walls imagining what might have happened between him and Natalie.

'Sal?' he says as he walks into the living room where I'm sat, in silence, with only the small table lamp on. 'What are you doing?'

'Nothing,' I say.

He walks around to stand in front of me, then gets down on his knees so his face is level with mine. His lovely, lovely face. Looking at him, I can't believe the sort of thoughts I've been having the last two hours. Where he, a police officer, got into a relationship with a teenage girl then lied and covered it up for years. No. There's no way. No way at all.

He smiles, which causes the left side of his cheek to form a dimple, and I bury my face in his shoulder as he wraps me in a tight hug.

'You're nuts,' he says as I breathe in his smell; late night coffee and fresh air. He pulls back and studies me in the soft light of the lamp. 'Everything alright?'

I nod. 'You're late,' I say.

He sighs, blows out his cheeks. 'Been a shit one today. I'm back in at seven, too. None of this "stress leave" malarkey for me,' he teases.

'Well actually, none of it for me soon either.'

'What do you mean?' He comes and sits next to me on the sofa, pulling me back so I'm sat in the nook of his arm.

'I saw your mum earlier. She talked to Holden and got me back on shift from Monday.' I look up at him as he raises his eyebrows.

'Wow,' he says. 'Still meddling then, even after she's retired.'

I can't tell if he's annoyed or impressed at the influence his mum still has over the force, so I simply shrug.

'You don't seem that pleased?' he adds.

'I am,' I say, wriggling out of his grasp.

'You were dying to go back a few days ago. What's happened? You got used to the lady of leisure lifestyle? Pjs until midday? Bit of *Loose Women* to liven things up?' He laughs and I try to laugh back but his assumption of how my days have been spent recently couldn't be further from the truth and it makes me uneasy, how quickly I've formed a life Nath knows nothing about. Shows how easy it is, I guess.

'I'm just tired,' I say. 'There's a lot going on with the wedding and everything...'

'Ah yes,' he says, standing up. 'The wedding. Can't believe in just over one week's time you'll be my actual legal property.' He smiles and reaches out his hand to pull me off the sofa then holds his finger up to say 'wait there' as he disappears from the room.

The Bluetooth speaker in the corner of the room lights up and a second later I hear the familiar opening notes of our first dance song. Nath appears in the doorway, a smile on his face and says, 'May I have this dance?'

I let myself be pulled into the centre of the room where he places one hand around my waist, one intertwined with my hand. He sways me gently around the room, his voice singing along to the song, deep and slightly off key but familiar and warm and safe. We stay like that until the song finishes, and for a moment longer until Nath tips my face up to his and kisses me.

'I love you,' he says, a phrase he must have said a thousand times by now – a phrase that used to be so simple and obvious but now feels brittle.

'Sal, what's wrong?' he says, stepping back, his eyebrows knotted in a frown as he studies me. 'Is it the wedding?'

I shake my head, eyes down.

'Look,' he says. 'I know there's something going on. I was talking to Mum about it yesterday.'

I must show something like panic on my face at this.

'Sorry, I know you like to keep things private but she's my mum, and she's basically your mum too.' He laughs, the familiar joke on the tip of his tongue about how I went from feeling like his little sister to something very, very different. 'She said you might be experiencing some grief again,' he says, no longer smiling.

'Grief?'

'About your dad not being here to walk you up the aisle.'

'Oh,' I say.

'I should have thought about it before,' Nath carries on, sure he's on the right path now. 'Of course it's going to bring that back up for you; neither of your parents are around to see you get married and that must be hard, Sal.

I can't imagine. But we're going to be our own family, you and me. You know your dad will be watching us – from somewhere! – and he'd be so happy, Sal. You know how much he loved the idea of us being together,' Nath says, smiling like it's that simple.

But his statement has only made things more complicated because he's right. There was a point where Dad *did* want me to be with Nathan. But something made him change his mind. Which means maybe he knew about something back then that I didn't.

I need to find out what it was.

CHAPTER TWENTY-SIX

The next day, I'm dripping with sweat, surrounded by my dad's things in the attic. Though it's a long-shot, I'm looking for evidence that Dad knew something bad about Nathan – that he had a reason for not wanting us to be together.

It's absolutely roasting up here and my whole body itches as I sit surrounded by fibreglass. So far, I've found nothing of interest. I've had to force down the grief bubbling in the pit of my stomach with every single thing I unearth from these boxes. The blue-and-red patchwork dressing gown Dad loved wearing to read the paper on a Sunday morning; the pile of Mick Herron books with their corners neatly folded down to mark his progress; the once starched white shirt he wore every day for thirty years as a police officer now creased where it's sat folded for six months since his death. My dad, in boxes.

So far, I've found nothing at all to suggest that Nathan was up to something dodgy with Natalie, and nothing to give my dad a reason to warn me off marrying him, either.

Instead, I've lost myself in the past; allowed the grief of losing him to finally take over. My eyes glisten as I find an old photo album – it's orange with a green dragon on the front and I instantly recognise it from when I was a little girl. I had forgotten this existed but the memory of carefully placing the photographs behind the cellophane pages comes back to me instantly; I can almost smell the scent of my mum's perfume and the feel of her warm skin as I sat on her lap in front of the gas fire in our old living room and put the album together.

But that memory, that scent, probably isn't even real. Mum left when I was so small I can't be sure any of my memories aren't simply something I invented in my head based on films of what a mother should be like. She was never a real mother.

As I flick through the photograph album, reliving the childhood memories I've not let myself look at for years, it's more obvious than ever my mum was not part of my life and the only people who have been consistently there for me are Dad, Amanda and Nathan. I watch us grow up together through the pages. The two of us as toddlers in the blue paddling pool my dad used to complain endlessly about blowing up for us in summer; Nath in Spider-Man trunks and me in a bright pink swimming costume, smiling with my two front teeth missing.

The next section is of holidays. I was an organised child; I smile to myself. Meticulous. I remember telling my mum how I was going to make sure each album section had a theme as she nodded seriously, telling me what a good idea that was.

Here we are in Spain when I'm still a toddler, bare-chested with a thick padded nappy. A couple of years later, the five of us – me, Nath, Mum, Dad and Amanda – are on a beach I don't recognise but think might have been somewhere in Devon. People always mistook us for one big family – with Nath and I as siblings. There was a time I wondered whether Dad and Amanda would turn into something romantic, but it never did.

I turn the page and here we are again, a happy foursome, on another holiday. This time, though, I recognise the beach. Because I was standing on it just yesterday.

I study it – we did go to Weston a lot when I was young. It was the closest beach to us, the only one we could make as a – rather long – day trip. In this picture, the Grand Pier is just behind us in the distance and I'm pointing at a donkey with a huge smile on my face. Dad has me pulled into him closely and we look happy. Really, really happy. The more I look at the photo, the more I remember that day. We drove down early one weekend morning, so early that Dad carried me out of my bed still wrapped in my duvet then placed me in the car still tucked up. *Snug as a bug in a rug*, he used to say.

There's something else he used to say that's coming back to me now. Something about Weston.

It's like the land that time forgot down here. It would make the perfect place to disappear.

I press the photo album shut, the memory of his words unfurling something uncomfortable in me as I move onto another set of photos.

They're all pictures of my mum. God, she was beautiful. And so young. I've never considered how young she was when she left – she can't even have been thirty. I stroke the cellophane covering of the photo as I flick past her face over the years. There's one where I'm cuddled up in a blanket as a newborn, close to her chest. She's skinny – too skinny for someone who has recently given birth. She doesn't look well, but I can't deny the look of love on her drawn face as she stares down at the bundle in her arms – me. I turn the page and there's that same look a few years later – just before she left. How could someone who looked at me like that leave me like they did?

I think of Natalie, how desperate she seemed for information about the baby she abandoned. Did my mum ever feel that way about me? It doesn't make any sense – not for her, or for Natalie. If you want to be with someone, you don't disappear from their lives. Yet both Natalie and my mum did exactly that.

I go back to the Weston photo. Natalie said someone had forced her to go into hiding, that she never wanted to leave her daughter. Who would want Natalie to hide away like that? The nagging feeling that Nathan is tied up in this – and my dad somehow knew – claws at the back of my skull.

Around the attic, there are only five unopened boxes left. I wipe the sweat from my brow and line them up side by side, leaving the biggest one to last. I need to keep looking.

The first is crap. Old books, CDs and DVDs. The second no more use – t-shirts, shorts, holiday clothes. The third the same, the fourth is crossword books and old electrical manuals. Without stopping for sentimentality, the exercise is a lot quicker and soon the fifth and final box is in front of me. I pull it towards me, it's heavy. I lift the lid – paperwork. I take a stack of paper off the top and ready myself for the task of going through it line by line. A lot of what is in front of me is financial information and I have a vivid memory of Dad telling me as a little girl that if you want to catch the bad guys, you should always follow the money. Our teatime conversations were unusual that way; Dad never treated me like a child, more like a detective in training. That I never followed him into detective work always felt like a failure, though he never made me feel that way. Dad was proud of me – of my job.

He knew how much I loved helping people, being there for them in the worst moments of their lives.

Dad always said that most of a detective's job was boring, painstaking work. Things like going line by line through financial data. I ignore my sense of disloyalty to my dad as I flip through his bank statements to get to the year I'm looking for – the year Shane was killed, the year Natalie disappeared. When I find the correct statement, I slow down, running my finger slowly down every single line of transactions to make sure I don't miss a thing.

And that's how I find it – a payment. Five thousand pounds withdrawn by my dad the day after Shane Blackwell was murdered.

CHAPTER TWENTY-SEVEN

I sit at the kitchen table with the bank statement in my hand. Five thousand pounds withdrawn from my dad's bank account. I try to think of scenarios where this isn't anything to do with Shane Blackwell's murder and Natalie's subsequent disappearance. My dad wasn't flash with cash; he wouldn't randomly withdraw that amount of money for no reason. Did he buy anything significant around then? I don't remember any big purchases, no new cars or anything like that – perhaps he intended to spend it and never did. But if so, what happened to all that money?

We were never badly off; we never struggled for money. But Dad was careful. He grew up poor – three kids in a two-up-two-down council house in Coventry. His police pay wasn't anything to complain about and he lived well within his means – drove his old car until he died and paid the mortgage off in full rather than get a bigger house. He always said he wanted to make sure he left me with something – and he did. Not a fortune, but enough that Nath and I are pretty comfortable. So why would my dad need five grand in cash? And what has it got to do with why he didn't want me to be with Nathan?

The sound of my phone ringing from upstairs temporarily puts a stop to my thoughts as I go to retrieve it. It's Amanda.

'Hi Sal,' she says. 'Everything okay?'

'Yes, good,' I say, twiddling my hair around my finger.

There's silence down the line for a moment and I almost fill it but can't think of anything I could say that wouldn't immediately alert Amanda to my state of crisis.

'Okay,' she says, finally. 'Well, that was it really. Nath working today?'

'Yes,' I say, and I can hear how stiff I sound but can't shake it.

'What did he have to say about your car getting stolen?' she asks.

I sigh. 'I haven't had a chance to tell him yet,' I say and for a moment Amanda says nothing, so I add, 'I will.'

'Right,' she says, as if it's totally normal behaviour for me to keep this from Nathan. 'Look, shall I pop over?' she asks. 'Chloe is coming over after school with Abigail but until then I've got nothing on.' The reminder of the huge breach of trust I've caused by telling Natalie about Abigail makes me feel sick again. 'Sally?'

'Sorry,' I say. 'Signal's bad today...' It's a stupid lie because why would my signal be bad? But Amanda doesn't pull me up on it.

'Don't worry if you're busy,' she says instead. 'Or just want a day on your own.'

'Yes, sorry. I'm doing some wedding things... Last-minute stuff, you know.'

''Course,' she says, kindly and I imagine asking her about my dad. About the money, about Nathan. But I can't. What if she doesn't know? What if I cause her the same level of terror as I'm currently going through?

Or worse than that, what if she does know and it turns out that every single significant person in my life has been keeping something from me for years?

Once Amanda has rung off, and agreed to leave me alone for the day, I go back to staring at the bank statement uselessly. I've been telling myself I have no idea what my dad might have used this money for but deep down, that's not true. Natalie said people had a vested interest in making sure she stayed disappeared. Someone like her couldn't disappear on her own; she'd need money. It never made sense to me how easily she disappeared after she killed Shane. It wasn't like she was some sort of criminal genius; she dropped out of high school without any GCSEs and stayed in her mum's council flat basically being Shane's drug mule, or sex worker depending on what he fancied at the time. She had no one but Shane. I have never let myself think too deeply about how she stayed hidden for so long — after all, my own mum managed it perfectly well — but now it makes sense that someone else was involved. But do I honestly believe my dad withdrew that cash to help Natalie disappear?

I try and remember what Natalie said about Alex and using her body to pay her rent. Was it a recent thing? Or going back years? If my dad helped her to get away initially, there's no sign he continued to give her money. But my dad wasn't a stupid man and would likely have another account for this purpose. Regardless, if he was helping her financially, it must have stopped after his death. This is insane. There's no proof the money withdrawn that day has anything to do with Natalie. And no proof that my lovely, safe, dependable dad would ever do something so bizarre, so utterly corrupt. But the nagging feeling that Natalie, Nathan, and my dad are all tied together somehow won't go away, and I know I can't simply forget about this.

There's only one person I can think of that might be able to give me some answers.

I take out my phone and dial her number. She doesn't answer. I ring again, and again, and again.

Eventually, the line connects.

'Yoooo,' she says down the line, her voice barely audible above the banging of the bass in the background.

'Tiff,' I say. 'Where are you?'

'Arcade,' she says as I hear the background sounds get gradually quieter. When the booming bass is a mere backdrop, she says, 'What you ringing me for? Why don't you WhatsApp like a normal person?'

Despite everything going on, I laugh. 'Normal people do make phone calls, you know.'

She makes a 'tssk' sound and I imagine her on the busy high street in Weston on a Friday afternoon.

'I need a favour,' I say.

'Another one?'

'Yes.'

She's quiet for a minute like she's thinking about it. 'Well, go on then! What is it?'

'I need you to ask Layla if she recognises someone for me.'

'Mate, I barely know her!'

'Please,' I say, hearing the desperation in my voice. 'Please, there's no one else I can ask. I need to know if she ever saw this man with Sally.'

'Jeez, you're obsessed,' Tiff says. 'But whatever. Send me a pic. I'll send it to her.'

I say thank you, but she's already gone.

I flick through pictures of my dad, my lovely, lovely dad, feeling sick, then hit send on WhatsApp. I tap the screen anxiously, expecting Tiff to reply straight away but minutes go by and there's nothing. I sit like that for another half hour before giving up and going back to studying my dad's bank statement, hoping for more answers, but find nothing.

By the time my phone beeps with a message, I'm completely insane with waiting, so when Tiff's name appears on the screen I jump on it. If Layla has told her she does know my dad, that's it. Isn't it? There will be no way I can look past this; no way of coming back.

I open the message.

> Laylas never seen that man in her life!

The relief is so overpowering I actually cry, then gather myself and before I can think better of it, send another photo. This time, it's of Nathan, with the question: *How about him?*

The reply comes back within seconds.

> She says no but is that your man? He's fit!

Whether it's from relief or despair, I don't know, but I laugh.

CHAPTER TWENTY-EIGHT

Monday morning comes quickly and I'm back on shift like I've never been away – the only proof anything out of the ordinary has happened in the last few weeks is the rental car I had to drive on the way in. My return is greeted with barely more than an *alright?* from my colleagues and a nod of the head from Inspector Holden as she passes me in the bunker on the way to get her coffee.

For the last week, I've been propelled forward by the idea I could help Sally – *Natalie* – if I just kept trying. But where did it get me? I started questioning my own dad's integrity. A man who never did a single thing to make me doubt him. Tiff said Layla has never seen him, or Nath, at Natalie's flat. If they were wrapped up with her somehow, they would have surely gone to the flat at some point in the last five years. The money my dad withdrew must be a coincidence. That's all. Yes, it sounds like Nathan definitely crossed a line with Natalie when he was undercover, and yes, he never told me about it. But so what? It's in the past. Natalie doesn't want, or need, my help – no one is asking me to burn down my whole life for the sake of something that is in the past. So, I'm letting it go. As much as possible, I'm going to carry on as normal.

The good thing about my job is that it's the perfect place to bury your own troubles. There's no chance to worry about anything much except the endless stream of calls that come in. It's not the busiest day today, Mondays rarely are, but there's enough going on across the patch to keep me concentrating on the task at hand and before I realise it, it's almost time for me to leave. I'm finishing up a call with Betty – one of our regular callers whose son is an absolute waste of space. She's seventy-four but seems more like ninety and he's a heroin addict, always coming around to her small terraced house, ransacking it for cash and leaving her terrified.

'Alright, love. They're on their way,' I tell her once I've confirmed two uniforms are nearby.

'I'm so sorry about this,' she says, her voice small and embarrassed, like it always is. 'I wouldn't call but he's not alone today.'

'You don't need to apologise. You just stay locked in the bathroom and wait for the police to arrive, okay?' I type up the notes on the incident log while Betty stays on the line. People think victims and criminals don't know each other – it's always a stranger you imagine being a threat to you, isn't it? A man comes in your house and threatens you with a knife for your money; you don't imagine the hand holding the weapon might be your son's. But it's so often the case. Still, I'm not hugely worried about Betty. As long as she keeps that door shut, her son will soon find some cash lying around and disappear once more until we pick him up again. He'll likely be out again soon as the CPS will say he's not a real threat or something stupid like that, like they always do, and the cycle will continue. Betty will be terrified, and we'll be her son's taxi service to somewhere else.

All I can do is help when she calls.

'Betty?' I say. 'Two police officers are walking up the steps to your front door right now.'

'Oh, good,' she says. 'I'm so sorry.'

'You stay on the line until I say so, okay?' I watch the bodycam footage of the two officers I've deployed to the scene as they enter Betty's house. It's empty, her son already long gone by the looks of it. They confirm the fact and say they'll take over, so I say bye to Betty and hang up the call, then finish typing up the incident notes and start to log out of the system.

I rub my eyes, the familiar adrenaline from working on a live job coursing through me. Nath is off shift tonight, finally our schedules are aligning, and he's said he'll cook dinner. In less than an hour, I'll be back in my house, with my husband-to-be, a glass of wine in hand and, most likely, a huge bowl of pasta waiting for me. Maybe that's enough — maybe this is enough. I don't need to know what happened with Natalie; Natalie is like Betty's son. She is the bad guy. The threat. The thing to hide away from. She is not Betty; she doesn't need saving.

'Heading off?' Lara asks next to me. Lara's new, only in her early twenties and she still has that look of terror every time the phone rings. They've sat her next to me to train her up, but I've not done the best job today, basically ignoring her in favour of getting on with my own work.

'Yeah, shouldn't you be finishing too?' I ask as I pull my jacket from the back of my chair.

'I'm signing out in a sec,' she says. 'Carrie's running late,' she adds in a whisper.

I sigh. Carrie is from E shift, the one that's supposed to take over after us and she's notoriously shit at turning up on time but she's been with the force for twenty-odd years so we end up covering for her sometimes.

'Don't stay too long,' I say as I pack up the rest of my stuff. Lara smiles as she presses to take another call.

There are three coffee mugs on my station, and I groan at the thought of washing them up before I leave. I know I have to; it's not fair on Ed who will take over after me. There's nothing more infuriating than when you start your shift surrounded by the debris of the person before you. I'll take them on the way out.

I call *bye* to the rest of the shift and mouth it to Lara, who's still on a call, but she shakes her head, her eyes wide, and mouths 'wait' back at me. I inwardly groan.

'Okay, so what time was she actually picked up?' Lara says, typing notes into the incident log as the person on the other end replies. 'And can I have a description of the woman who picked her up, please?'

Lara starts gesticulating at me wildly to listen in to the call, panic in her eyes. She's new; they're all like this at first. Every call is life or death when it hardly ever is. But still, I pick up the other headset and Lara mouths 'thank you' to me.

'...I didn't know, I didn't realise she wasn't the sister, she said she was—'

'Yes,' Lara says, cutting in. 'I understand that. But please, I need a description of the woman who took the child.'

My eyebrows shoot up. Perhaps Lara wasn't being dramatic after all.

As the caller starts talking again, I scroll up the incident log as Lara types beside me. A child is being reported missing after being picked up from her after-school club today by a woman claiming to be her aunt.

'...She wasn't very old, maybe mid-twenties. Dark hair, about my height—'

'What's your height?' Lara says.

'Five-foot-four. She might be a bit shorter. I'm sorry, I don't know.' In the background of the call, I hear a woman, presumably the mother, crying.

'Okay, I'm sending officers out to you immediately. Can I please speak to Abigail's mother now?' Lara says and I stop breathing for a second. Scroll back up the log, reading.

> ABIGAIL LAING ABDUCTED FROM SCHOOL APPROX 16:45
>
> MOTHER CHLOE LAING DUE TO PICK HER UP AT 18:00
>
> MOTHER ON SCENE WITH CALLER, JANE THOMPSON, TEACHER

All I can hear is the beating of my own heart in my ears, pure panic spreading through me as everything gets hot. Then, I tune back into the call. Chloe is crying as Lara tries to calm her down to get more details from her. What was Abigail wearing this morning when she went to school? A blue denim dress, white frilly socks and a yellow hair band. Where is Abigail's dad right now? On his way to the school from work. Does Chloe recognise the description of the woman who took Abigail? No. Can Chloe think of anyone who might want to take Abigail for any reason? Indecipherable cries.

Lara looks to me for reassurance that she's asking all the right questions and I nod, like I'm in any way in control of this situation. Like I'm not the one who caused it.

Abigail's been taken.

Natalie has taken her.

CHAPTER TWENTY-NINE

I listen into the call for long enough to know I'm not going to gain any more useful information from Chloe or the teacher who handed Abigail over to Natalie. The description she gave fits Natalie – and as yet, no one has made the link that the woman who has taken Abigail is actually her birth mother. As soon as Carrie from the next shift arrives, I make my excuses and leave. I'm already in deep shit for not declaring my connection to Chloe while listening in on the call and I'm sure that will soon come out, but for now, I need to get away.

In the car, I text Nathan and tell him to get home straight away – it's an emergency.

I do the twenty-minute drive to our house in fifteen minutes flat and rush into the house.

'Sal?' Nath meets me at the door, his brow knotted in a tight frown. 'What's happened?' He's still in his work clothes, a sharp white shirt and black trousers, but his top button has been loosened.

'It's Abigail, she's been taken.'

'What?'

I push past him to get into the house and he shuts the front door behind me, following me into the kitchen.

'Did that just happen on shift?' he asks.

I nod, and he gets out his phone, but I grab it from him.

'What are you doing? I need to call Mum.'

'No,' I say. 'You can't.'

'Why not? What's going on?'

'Please, sit down. I need to... I need to explain...'

'Okay,' Nath says, taking a seat but his leg jitters beneath the table. 'Tell me what happened. Did you take the call?'

I shake my head. 'No, Lara did.'

'So, what's happened?'

I drum my fingers against the table, wondering how much to tell him. But it's too late for secrets now, I need to tell him everything. Look where lying has got me.

'I was just about to finish my shift, but Lara gets the call. Abi was at an after-school club. Chloe was due to pick her up at six like she does every Monday but a woman arrived around four forty-five p.m. and said she was Abigail's aunt—'

'Doesn't the school have a protocol in place for things like that? You can't just rock up and take a child—'

'I don't know. I don't know how it happened, but it has.'

'Fucking hell,' Nath says, rubbing his face. 'Chloe must be beside herself. So, what's happening? Why did you leave? I thought you'd want to stay and help. Do they have any idea who this woman is?'

I shake my head. 'No, they don't. But Nath, *I* do...'

He cocks his head to the side, waiting for a beat. For me to say more but I don't – I can't.

'I don't understand,' he eventually says. 'If you know who took her, why didn't you tell Lara and get her found?'

'Because I needed to talk to you first.'

Another beat. A moment where if he didn't have anything to do with Natalie the silence would be filled with questions. But it's not, and that's when I know for sure I was right about Nathan and Natalie.

'It's Natalie Pierce,' I say.

Nathan's jaw clenches. He gulps, the Adam's apple at his throat jumping, then recovers. 'Abigail's birth mum,' he says.

I nod.

'You think Natalie came back and took her?'

'Yes, I do.'

He nods, stands up and walks around the kitchen. 'You need to tell the police that, Sal.'

The atmosphere between us hardens; Nathan's shoulders tense up as he turns his back to me. There should be a sense of urgency in our conversation but there's not; it's a stand-off. Both of us unwilling to admit what we know until we understand how much the other does.

He turns around, his dark brown eyes meeting mine. The eyes I've looked into for almost as many years as my own – the eyes I've known for my whole life. But do I truly know this man at all?

'Sal...' His hands push his hair back from his face as he shakes his head.

'I'm scared what they're going to find out...' I say and his face switches.

'What?' he asks, his expression riddled with confusion.

I can't say anything, the words stick in my throat.

He shakes his head, his eyes wide under a knotted brow. 'What do you mean? Find out about what? I don't understand, Sal.'

'About why Natalie went missing. About what happened back then,' I say, watching his face for clues.

Nath looks at me like I'm insane. 'She disappeared because she killed Shane—'

'She says she didn't—'

'What?' Nath steps towards me, his hands reaching for me, but I flinch away. 'You *talked* to her?'

I nod.

'When?'

'Stop questioning me!' I throw my hands up like a petulant child and Nath sits down, his hands up in defeat. 'You know Natalie. Don't you? You met her when you were undercover.'

I train my eyes on his as he opens and shuts his mouth like a fish. Then, slowly, he nods. 'Yes. I knew Natalie.'

Though I knew it to be true, the admission is a gut punch. For a second, I can't speak as I look at the man I thought I knew better than anyone becoming a stranger in front of my eyes.

'What happened between you?'

'Nothing,' Nath says. 'Jesus, Sally. What do you mean?'

I shake my head. 'Don't look at me like that. You're the liar—'

'It was a job! You know I would never be able to tell you about working undercover—'

'But you weren't just undercover, were you? Something happened. Between you and Natalie. Didn't it?'

Nath shakes his head, his eyes sad and hopeless. 'No. Nothing happened between me and Natalie – nothing like *that*. I swear to you, Sal. I would never do that.'

I'm torn. I want to believe him so badly. Because the alternative is too awful to bear.

'Sal, please. You said you think she's taken Abigail. How do you know that?'

I bite my lip because now it's my turn for a confession. 'Because I might have told her where Abigail was.'

'Fuck,' he says. 'What are we going to do?'

CHAPTER THIRTY

It takes less than ten minutes to explain to Nathan the full, convoluted story of how I know Natalie – the years of crossed emails, the suspicion something was wrong, the insane trips I've made to try and find her. And then, we get onto Alex.

'Sorry, what?' Nathan says. 'This bloke turned up in the woods?'

I nod.

'Why was he there?'

'I don't know.'

Nath scratches his head, and I can see his brain whirring.

'So what happened then?' he asks and I look away. 'Sal, what are you not telling me?'

'I needed Natalie to trust me, okay? When she saw someone drive up, she was furious at me. Thought I'd called someone to come out…'

'Right, and…?'

'I drove off. Fast. I didn't see Alex step in front of the car until it was too late.'

Silence.

I look up. Nath's neck is pumping, a fat purple vein beating in time with the ticking of the kitchen clock.

'Are you saying… Are you saying you hit him?'

I nod.

Nath nods back, once. He's gone still and I see this is how he must be at work – poised and unemotional in circumstances that would make anyone else crumble.

'Did anyone see you?'

I shake my head. 'I don't think so.'

'Your car,' he says, his eyes narrowing. 'It wasn't stolen.'

I shake my head. 'I pushed it off a cliff to cover the damage...'

He opens his mouth while taking this in, then recovers. 'The man you hit? Is he...'

'He's fine. I think. Natalie went to see him in the hospital, and he isn't going to say anything—'

He takes a step back, as if this is the most shocking thing that I've told him so far. 'Natalie went to... Fucking hell, Sal. Are you stupid? What about CCTV?'

'She wore a cap,' I say feebly.

He shakes his head and then closes his eyes, breathes in heavily and refocuses.

'So then what happened?'

'After she came out of hospital, she was different. We argued—'

'What about?'

'I don't know. It was like she'd changed her mind about me, she had my phone...' I look down at my phone, my face unlocking it. A picture of me and my dad smiles back at me. 'Oh my god,' I say. I hold the screen up to Nath.

'What?'

'My dad,' I say. 'She saw my dad.'

Nath shakes his head. 'I don't understand—'

'She must have seen the picture of my dad. I told you something changed. Before that, she was learning to trust me. Then out of nowhere, she changed – it's this.' I hold the screen up again, waving it in his face. 'My dad. She knew my dad. The day after Shane was killed, my dad withdrew a chunk of money. I think he helped Natalie to disappear. She must have thought I knew – that I was in on it.'

'In on what?' Nath says.

'I don't know!' I fall back in my chair, exhausted. Every time I get closer to understanding what's going on, I realise how fucking little I know. 'But that doesn't matter right now. What matters is that she wanted information about Abigail – Grace, she calls her – and I gave it to her—'

'Why would you do that?'

'Because she said if I told her about Abi, she'd tell me what she knew about you!'

Nath has the decency to look ashamed – this isn't my mess, it's his.

'She told me she loved you,' I say and at this his face contorts and he stands up, planting his hands on the table.

'It wasn't like that…' he says, then walks around the table to kneel in front of me. He takes my hands in his, gripping them tightly. 'I promise you, Sal. Whatever you're thinking happened between Natalie and I, you're wrong.'

A tear falls down my cheek before I can stop it and he reaches to brush it away. I let him; the touch of his hands on my cheek feels so familiar, it's all I want.

'What are we going to do?' I ask. 'If we tell the police Natalie has Abigail, I'll have to tell them everything I did – I helped a known criminal stay hidden, I *ran over* a man, I led Natalie straight to Abigail.' As I say the list out loud, the magnitude of my situation hits me.

I could go to prison for this.

I'll certainly lose my job at the very least.

But it's bigger than just me. They'll investigate my connection with Natalie. Whatever my dad did, whatever his involvement – and I'm sure there is some involvement – it will come out. His name, his legacy with the police, will be totally obliterated. And Nathan? Do I really trust him when he tells me nothing happened between them?

'We need to find her,' Nathan says. 'Before the police do – we find Abigail and bring her back. If we find her first, we'll stand some chance of controlling what happens next.'

I nod and wait for Nath to continue, as if he is going to have formed an entire plan of *how* we go about finding a missing child and a woman who's stayed hidden successfully for five years.

—

Half an hour later, we're round the kitchen table with notepaper strewn between the two of us as we work out our plan. Nathan made a call into work – they can't have him on the investigation team because of his personal connection with Abi but they did let him know they were searching for Natalie. Once Chloe revealed exactly who Abigail was and her ties to Natalie, and the officers realised the description of the abductor matched what Natalie might look like now, all of their efforts went into finding her. Every officer has been called in to help with the search, but it's not yet become public news. They're hoping to find her before they do a national appeal. But they won't put it off for long; if she's not found within the hour, I expect Natalie and Abigail's faces will be plastered over every news outlet across the country.

Amanda has been on the phone – she is at Chloe's, trying to help her navigate the police involvement.

I'm so relieved the police have put together that it's Natalie who took Abigail. It takes some of the guilt away from me for not telling them what's going on. Though I know if I were honest with them about the last few days, they'd have more chance of finding her. But still, I have Nathan – I trust him to find her. I trust us both.

'So, Natalie doesn't have a car—'

'That I know of, anyway.'

'So she's likely to be on foot and public transport,' Nath says and I nod. 'She doesn't have access to a lot of cash—'

'Again, not that I know of.'

'But she does know people that have access to places to hide.'

'You don't think she'd be back in the woods where I found her before?' I ask hopefully, as if it will be this simple.

But Nath shakes his head. 'She's not stupid, she'll know you'd go straight there.' He taps his pen against the paper. 'Did she tell you about anyone else in her life? Friends? Boyfriends?'

'She has one friend,' I say, relieved to finally have something useful to add. 'Layla.'

'Layla what?'

I shake my head. 'I don't know but she lives in the same apartment building as Natalie.' I tell Nath the name of the building and he gets out his work laptop then logs into GENIE, the police crime database we have access to. I frown as I watch him type in the address, knowing if this ever gets audited, Nath will have to explain why he was looking up Layla's address to begin with. But in the scheme of things, this is a minor crime.

'Layla Monroe,' Nath says, turning his laptop to face me. 'Previous charges for ASB and drugs, nothing major.'

I study the mugshot in front of me – Layla looks much the same as most young women we have on GENIE. Grey-skinned and dead behind the eyes.

'Does Layla know about you?' Nath asks me and I wrinkle my nose.

'Kind of,' I say.

'What does that mean?'

'I met this girl,' I say. 'You remember I told you about the girl on the beach? She's called Tiff.'

Nathan arches his eyebrows and nods.

'She came up to the flat with me. I shouldn't have got her involved I know. Anyway, she found Layla and the other day, I…'

'You what?'

'I asked Tiff to show Layla some photos, to see if she recognised them from Natalie's flat…'

'Photos of who?'

'My dad,' I say.

'And who else?'

I sigh and look away. 'You.'

Nath doesn't say anything.

'Aren't you going to ask me what Layla said?'

'I don't need to,' he says, his voice flat. 'At least, not about me.'

'She didn't recognise my dad either,' I say quietly, and he nods.

'We should find out if Layla knows anything about Natalie's plans,' he says. 'Can you speak to this Tiff again?'

I go into the other room to make the call and this time Tiff answers on the first ring. Today the line is quieter, and I imagine her in her tidy small living room looking after her brothers and sisters. I explain something bad has happened and I need her to go and speak to Layla, to take the phone up to Layla's flat. She tuts at me but then does what I say as I wait for a couple of minutes while Tiff explains to Layla what's happening.

I go back into the kitchen and put the phone on speaker so Nath can listen in.

'Hello?' Layla comes on the line sounding wary.

'Layla?' I say. 'My name's Sally. I think Tiff told you about me…?'

'What do you want?'

'Look, something's happened. Natalie has done something—'

'I don't know no Natalie,' Layla says, and I realise my mistake.

'Sally, sorry. That's her real name — Sally's. She's not Sally Jones, her real name is Natalie. Natalie Pierce.'

There's rustling on the other end of the phone and I hear Tiff saying something but can't make out the words, then Layla comes back on the phone.

'I don't know where she is. I ain't seen her for over a week, alright? It's nothing to do with me.'

'Layla,' Nathan cuts in and I shake my head, but he holds his hand up. 'My name's Nathan, I'm a police officer but I'm not working for the police right now. They don't know about this phone call and for Natalie's — *Sally's* — sake, I'm going to keep it that way. But we need to find her before they do, do you understand?'

'The fuck you talking about? I've just told you I don't know where she is!'

'But do you know anything that might help us?'

'Why would I help you? I don't even know who you are.'

'In about an hour's time, you're going to see news reports of an abducted child here in Warwickshire. You're going to see a woman named Natalie Pierce all over the news – that's your friend Sally. She's a wanted murderer and today she went to her daughter's school and took her.'

'Shiiiit!' Tiff whistles down the phone and I imagine her eyes cartoon-wide as she hears this. 'That's mental! Like a film or summit.'

'We're worried about Natalie—' I start to say but Layla cuts in.

'Who's the kid?'

'What?'

'You said she abducted some kid. Sally wouldn't do nothing like that unless she had a good reason. So who is she?'

'It's Sally's daughter,' I say. 'She was adopted five years ago as a baby. Sally found out where she was, and she's taken her.'

'Well, good,' Layla says, sniffing. 'Social's always round here taking our kids and they can fuck off. Why shouldn't Sal get her baby back?'

I go to answer but Nath cuts me off. 'She won't get far with her daughter, Layla. The police will find her. And if they find her before we do, Natalie – Sally – will have no chance of ever seeing that little girl again. Do you understand? We're trying to help her.'

I hear Tiff talking to Layla but one of them has their hand over the phone so we can't make it out properly. Is she convincing Layla I'm a person she can trust? Or are they debating how to get rid of us both?

'Alright,' Layla says after a couple of minutes. 'I'll tell you what I know.'

CHAPTER THIRTY-ONE

Nath hangs up and we look at each other grimly.

Hearing Layla repeat what Natalie told her – that she was going to take her daughter 'back where she belonged' broke my heart. Natalie is so sure the police should never have taken Abigail in the first place, like she's the wronged party in all of this. What if she's right? I look at Nath, the man I've loved all my life, and wonder what, if any, part he's played in Natalie's downfall.

'Nath—'

'We'll talk, I promise,' he says, as if reading my mind. 'But Layla said Natalie is going to take Abigail to this address back in Weston.' He taps his phone where he's written down the address Layla hesitantly gave us. 'We need to be there when she arrives. We can still fix this.' He says it so resolutely, I almost believe him. I want to believe him, at least. And so, when he takes me by the hand and leads me out onto the street and into his car, I go, willingly – without asking him a thing.

-

This journey down the M5 is becoming all too familiar and for a while, Nath and I say nothing to each other as we watch the motorway speed by. I know the longer I leave it before I force Nath to answer my questions, the longer he'll have to come up with a story that paints him in a good light. But I can't bring myself to start asking because I know that once I do, there will be no going back.

'Sal,' he says, changing into the slower lane behind a large lorry. 'I know you must have questions. About Natalie. About, everything…'

He glances at me like a school kid who's been caught out and I can't help but feel a pang of love for him.

'I can't actually believe this is all happening,' I say, instead of *yes, I do, tell me everything* because I'm not ready to hear it, not yet.

'I know.'

We stop talking for a moment as a BMW angrily cuts in and out of our lane, as if we're holding him up. Nath mutters *tosser* as his cheeks flame.

Instead of thinking about the conversation we need to have I decide we should rehash everything Layla has told us.

'Layla said she gave Natalie five hundred pounds,' I say and Nath nods. 'Are you sure that's not enough for her to have hired a car?'

He frowns. 'She's not going to go to the hassle of hiring a car when she's in a rush. Does she even know how to drive?' he asks.

'Yes. She said she did,' I tell him, thinking back to the day we hit Alex and her offering to drive us out of Weston. 'But you're right, I doubt she's got her hands on a car.'

Nath nods. 'So it's likely she used public transport. Buses would take too long from Weston to Stratford and back,' he says, thinking out loud. 'Possibly the train, then.'

'Surely the police will already be tracking that,' I say.

Nath nods. 'It'll be the first thing they check, yeah – but it won't be easy to spot them on the cameras if Natalie is at all clever about it. They'd have got on the train just before rush hour and no one is going to look twice at a young woman and a schoolgirl at that time of day. Plus, she's probably got on and off to make sure no one can track her.'

I nod, though I'm not sure I can imagine Natalie being able to get Abigail to follow her commands for long enough to be hopping on and off trains.

'I don't know,' I say, aloud. 'Don't you remember when we took Abigail out that one day without Chloe and Matt? She was an absolute nightmare—'

Nath laughs. 'We had to pick her up in the end, didn't we? I'd forgotten about that,' he says, his smile disappearing as he thinks about what this could mean. 'If Abigail's making a fuss, then it's likely they'll get stopped on the train. Someone will notice them.'

I rub my eyes. 'Are we being stupid heading straight to this address?' I ask, suddenly unsure of our plan, which had felt so concrete a few minutes before. 'Would it be better to try and catch them at the station?'

Nath taps his fingers against the steering wheel, debating with himself.

'No,' he says after a moment, putting his foot on the accelerator and taking us back into the fast lane. 'We can't know for certain how she's travelling. All we know for sure is her end destination. Remember, the police don't know where she's going – only we do. We'll be able to get to her before any of this gets even messier than it already is.'

I nod, relieved to not be the one who has the final say in this.

I get out my phone and look at the address Layla gave us again. Despite Layla's earlier insistence she had no idea where Natalie had gone, after a bit of grilling she admitted Natalie had come to her the day before and begged her to help her out. Sally had told Layla she finally knew where her 'stolen' daughter was and all she needed was some cash from Layla to help get her back. The one thing Natalie didn't tell Layla was her real name – I guess she couldn't be sure even her best friend would help her if she knew who she really was. How sad.

'What if she changes her plans, though?' I say to Nath.

'She won't. She'll be there.'

I put the address into Google Maps and go on street view to get another look at the building Layla has told us Natalie is taking Abigail to. It's down a street not far from the high street in Weston, a basement flat underneath a tattoo parlour. I try and imagine what Natalie's long-term plan here is – surely she knows she'll never get away with hiding a young girl for very long? But then again, she doesn't know Abigail. She doesn't know about her fiery temper or how she'll have a total meltdown if she's not near her mum, Chloe, at all times. I wonder how Natalie got her to walk away from the school without Chloe in sight, but she is only five and if the teachers told her to do something, she would.

'I wasn't in a relationship with Natalie,' Nathan says, breaking me out of my thoughts. 'I know that's what you think.'

I sigh. 'She said she loved you.'

He shakes his head, but I notice the tips of his ears have turned red like they do when he's stressed – or angry. 'We became close when I was working undercover.'

'How close?' I say.

He turns to me, hurt written across his face. 'She was like a little sister to me.'

I nod, not wanting to say that's exactly what he used to call me – before he decided I was much more than that. If he recognises the phrase and knows what I'm thinking, he shows no sign of it, speeding the car up and undercutting another BMW.

'Nath,' I warn him. 'Be careful.'

'I want to tell you what happened,' he says, and I steel myself to hear the truth. But then, his phone rings, coming out over the car speakers. His mum's name shows on the dashboard.

'You should take it,' I say and so he does.

'Nathan?' Amanda's voice is terse.

'Hi Mum, you're on speaker. Sally's in the car,' he says, glancing at me.

There's a beat or two of silence. 'Does she know? About Abigail?'

Nath looks at me and I nod. 'Yes, Sally was on shift when the call came in. How's Chloe?'

'She's beside herself,' Amanda says. 'Where are you going?'

'What?'

'You said you were in the car. Where are you going? I thought you'd want to be on hand for Chloe.'

Nath looks at me, wide-eyed, and I wonder how he ever managed to survive undercover at all.

'We're out picking up supplies,' I say and Nath mouths 'what?' at me as I shrug.

'Supplies for what?'

'For Chloe and Matt…' I say rather stupidly. 'We thought it might be a long time, so we're going to get coffee and food and things they… might need…'

Amanda says nothing for a while. Then, 'I think all they need,' she says with a pause, 'is their daughter home.'

'Are you still at their house?' Nathan says. 'Is Matt there?'

'I'm here. Matt too,' Amanda replies. 'It's terrible, Nathan. Watching them both go through this. Truly terrible.'

I watch Nathan, wondering for the first time how much, if anything, Amanda knows about her own son's involvement in the Natalie Pierce case. Natalie's name hasn't been spoken but it's like a bad smell hovering on the line between us all.

'Call me when you're out of the car, Nathan. We need to talk, properly.'

'Yep, I will,' he says. 'Bye, Mum.' The call cuts off.

'Why does she want you to call her later?'

'No idea,' Nath says.

'Does she know? About you and Natalie?'

'For fuck's sake, Sal, I've told you there is no "me and Natalie".'

'You know what I mean.'

Nathan huffs. 'Mum doesn't know anything. And I'd like it to stay that way.'

'So why does she want you to call her back when I'm out of earshot?'

'She didn't say that. She said when I'm out of the car. You know she hates talking to people when they're driving; says it's not safe.'

I watch the side of his face as he concentrates on the road. The red tinge on the tips of his ears glows, but is this because he's lying or because we're in the most stressful situation of our entire lives?

'You're sure she doesn't know anything?' I ask and Nath turns to me, his face open and honest.

'No,' he says, his eyes boring into my mine. 'She doesn't know anything, I promise.'

I nod, relief flooding through me. It's one thing to think Nathan might have been keeping things from me but the idea of his mum – the closest thing I've got to a mum myself – being in on it is almost too much to bear.

I look at Google Maps on the car screen; we're only fifteen minutes away from the address.

'Do you think it's a coincidence Natalie and I met over email?' I ask.

Nath sighs. 'No,' he says. 'I'm sorry, Sal. I know you want to believe she's been your friend, but I think we have to admit this was probably Natalie's plan all along.'

'You think she used me to get to Abigail?'

'I think it's a strong possibility, yes.'

I furrow my brow. I know he's probably right; after all it's too big of a coincidence for she and I to have found each other otherwise, isn't it?

'But she never tried to find me,' I say. 'She didn't come looking for me. I went looking for her.'

'Maybe that's what she wanted,' he says.

I think back to turning up at the abandoned building Natalie was hiding out in. The look of surprise – and anger – on her face when she opened the door to me felt real, it still feels real. She didn't look pleased to see me. If she was using me to get to Abigail, surely she'd have wanted me to find her?

'It doesn't make sense,' I say.

'None of this makes sense,' Nath says.

I open my emails and look back at the last conversations Natalie and I had. Then, suddenly, it clicks.

'She ran away after I told her I was getting married,' I say, studying Nathan's reaction. 'When I told her I was changing my name. To yours.'

He flicks a glance to me, but then starts concentrating on the road again while I get out my phone and look for the email.

'I don't think she knew who I was before that,' I say. 'Not properly. She knew my name – she had a newspaper article about me in her flat about my Chief's Commendation – but I don't think she knew I was involved with you. Not until I sent this email.' I hold up my screen but Nath shakes his head, like he does whenever I try and show him something when he drives – though it's usually some silly TikTok that will make him laugh rather than accusatory information like this.

'She wasn't trying to find me, Nath,' I say. 'She was trying to run away from you.'

CHAPTER THIRTY-TWO

Nathan isn't talking to me. The idea that it's him Natalie is hiding from has made me re-evaluate everything I thought I knew. He seems hurt by the suggestion, but is he also worried that I've worked out the truth? But what is the truth? That he had a relationship with Natalie when he was undercover, she murdered Shane, got her baby taken away and went into hiding and years later is still running from Nathan? What could Nathan have possibly done to make her so afraid? The thing is, she didn't appear to be afraid of Nathan when I met her. She said it herself – she loved him.

Is there another reason finding out I was marrying him would send her running? I suppose maybe if she was still in love with him she might not want to stay in contact with the woman he was now marrying. But that doesn't explain the wrecked flat. She told me someone was warning her off – she said she was running from the police. Could she have simply meant Nathan? But until right now, Nathan didn't even know the two of us were in contact.

Google Maps tells us we're five minutes away. I look at my watch. If Natalie managed to get Abigail straight onto a train without any fuss, they could be here already. I quickly text Amanda.

> Any news on Abigail?

I watch as it says she's typing, then she stops. Starts again.

> No. Will update you if anything comes in.

At least that means they haven't caught them on the train before we've had a chance to talk to Natalie and get our stories straight. But it could also mean Natalie is somewhere far, far away with Abigail by now and we might never see either of them again.

No. I can't let myself think like that.

'Nath,' I say, breaking the silence between us. 'Before we go in, I need to know what happened between you and Natalie.'

He pulls over onto a residential street and turns off the engine. 'What do you think happened?' he says with hurt behind his words. 'What do you think I've done?'

I shake my head, tears threatening behind my eyes. 'I don't know. I really don't know. All I know is there's a young woman who says the police are working against her, someone is forcing her into hiding and she ran a mile when she found out I was marrying you. If you were me, what would you think?'

He sighs, rubbing his face. 'It looks bad, I know that. But you have to trust me when I say nothing ever happened between Natalie and me. Look, we're so close to finding her now – it's just around this corner. Can we please concentrate on getting Abigail back and then we'll talk? I promise.'

I nod, because he's right. The most important thing in all of this is Abigail's safety. Everything else can wait.

Nath starts the engine back up and we slowly drive around the corner onto the high street. The sound of seagulls and the roar of wind hits as I wind down my windows, letting the smell of chip fat and doughnuts into the car.

'Ready?' Nath says, turning off the car.

'Is this it?' I ask and he nods, getting out of the car, and I follow him.

I look up and down the grungy side street. It's pretty rough; there's graffiti all over the walls – not art, just scribbles and swear words. Lots of 'HARRY WOZ ERE' sort of things. There's a smell of stale fat and cigarettes and beneath our feet are hundreds of discarded fag ends. I try to imagine Abigail being led down here but I can't. She'd make a fuss – maybe she already has made a fuss. It's not safe, not the sort of place a child should be taken. Especially a child like Abigail who has been kept wrapped up in cotton wool by Chloe since the day she came home, protected from places like this – and from people like Natalie.

'It's up here,' Nathan says, pointing at the tattoo parlour I've seen on Google Maps.

I follow him as he strides up the street and rushes down the steps, banging heavily on the door of the basement flat.

'Natalie?' he shouts, banging on the door like he wants to break it. 'Open up. Natalie!'

After a few bashes, the shape of someone coming towards us through the criss-crossed dirty glass door appears. Is it her? Could it be this easy? What will we do when we find her? Grab Abigail and run? As the door swings open, it dawns on me how incredibly stupid we are because we have no plan for what happens once that door is open.

'The fuck are you banging for?'

But it's not her. A man — his age impossible to tell through a face of tattoos and a thick black beard — glares at Nathan with a cigarette hanging from his lips.

'We're looking for Natalie,' Nathan says, his voice not wavering despite the unexpected sight of this man blocking our way.

He sniffs. 'Not here, mate.' He goes to shut the door, but Nathan slams his foot in it, then pushes it open and shoves the man out of the way before going into the flat.

'Natalie! I know you're in here. Come out!' I follow him into the flat, while the tattooed man calmly stands by the door smoking his cigarette, shaking his head at us both.

'No idea who you're talking about, you nutters,' he says as I pass him.

The flat is small, two sparse rooms, and it's immediately apparent Natalie isn't here.

'Is there another way out?' Nath asks the man, who takes a big drag on his cigarette, blowing it in Nath's face, then shakes his head.

'You wanna tell me who this Natalie chick is, or you going to keep walking around my flat like you own the place?'

'Natalie Pierce,' I say, stepping in between the man and Nath, trying to de-escalate the situation. The man shakes his head.

'Never heard of her. Why do you think she's here?'

'Her friend gave us this address,' I say. 'Layla?'

The man laughs, a throaty ghastly sound that quickly descends into hacking coughs. 'Should have known it would have something to do with that mad bitch,' he says once he recovers.

'You know Layla?'

'My mental ex. She's always pulling shit like this.'

Nath storms out, anger radiating off him as I call Tiff, but she doesn't answer. A second passes and a text message comes up from a number I don't recognise.

> You really think I'd tell you where she is? Stupid fucking pigs. Hope you enjoyed your drive!

I show it to Nath who smashes his hand against the wall in anger.

'She fucking tricked us,' Nath says. 'And we fucking fell for it.'

'Where is she then? If she's not here, where the fuck is Natalie?'

Nathan looks at me, his eyes wide and scared, and I realise he has absolutely no idea.

PART FOUR

CHAPTER THIRTY-THREE

NATALIE

'Will Mummy be there?' Grace asks me again as I sit opposite her on the train.

'No, Gracie, remember—'

'My name's Abigail,' she says, in that strange little posh accent of hers and I try not to cry. She doesn't sound like my girl at all.

'Sorry, sweetie. I keep forgetting...' I turn to look out of the window as tall buildings flash past us. Grace has been good so far; she obviously hasn't been taught much about stranger danger. This is the thing when you bring up kids in them big posh houses, they don't learn not everyone in the world is their friend.

Except I am her friend. More than that, I'm her mum.

I haven't told her that, of course. Not yet. I'm sure she can sense it – we look so alike! How could she miss it? She has my big brown eyes and freckles on her little nose. But that voice, it's not mine. It's not hers either, not really. I'm sure she'll drop the posh accent after we spend more time together; she's probably putting it on, isn't she? To fit in with her posh fake parents. I wonder if I've lost my Midlands accent, being down south for so long, but I don't think so. I guess Gracie will learn to talk like me, after a while.

'When is Mummy coming?' she asks again, and I turn to face her then reach for her hand across the seat. She pulls back, her little face contorting into a frown.

'It's okay,' I tell her. 'Remember, we're going on an adventure. You remember? We're going to get fish and chips at the seaside.'

'I don't want fish and chips,' she says as a tear spills down her cheek. 'I want Mummy.'

The words dredge up some deep hurt in my stomach, like when we used to chuck stones into the river as kids and the mud and sludge and shit would rise to the top – it had always been there, this mulch, but we'd disturbed it. Brought it to the surface.

'Have some sweets,' I say, pulling out a bag of warmed Haribo from my backpack and waving them under her nose, like the bloody child catcher from that kid's film. Grace shakes her head and crosses her arms then turns away from me. I let it go; as long as she's quiet while she has her strop, that's all I need.

Alright, I admit it. I haven't exactly thought this plan through.

But when I saw that photo on Sally's phone outside the hospital – that man – I knew I had to get as far away from her as possible. I should have known she was caught up in all of this. She ain't safe – no matter how bloody nice she pretends to be. Only this time, I wouldn't be running away alone. Hearing about Grace – *Abigail* – as a real-life little girl, living where I grew up, it did something to me. It was like that river again, only the feelings weren't dark and murky – it was love. I've not let myself love Grace, not fully, not like a real mum, since I've been in hiding because I knew I couldn't. How can you let yourself love someone that you can't be with? The only place I've let Grace be alive was in those stupid emails to Sally. I invented my little girl and let us live out that perfect life through my words – it's all I thought I could ever have. A make-believe life with a make-believe daughter.

Only after meeting Sally, I realised it didn't have to be make-believe. Grace was real, she was real and out there and living a whole life without me. I just had to find her.

Once Sally told me her name and said she saw her all the time, it wasn't hard to find her online. I guessed she must live near Sally and I know Sally's address from her emails. Plus, Grace's stupid fake mum, *Chloe*, does exactly the sort of stupid thing you'd expect – sticks everything out in the open for everyone to see – as if she ain't scared of nothing. She's been bloody thrilled to snap endless photos of my little girl and stick them all over the internet. Imagine that! Putting all them pics up there for anyone to see. She covers Grace's face with that big stupid yellow smiling cartoon face, as if that protects her, but I knew it was her. I knew it was my girl. It only took finding Sally, who barely uses social media, then clicking through all her friends – she only has seventy-eight, which is a bit weird, isn't it, as most people have hundreds, a bit sad, really – until I found the profile for Chloe. After that, it didn't take me long to find Grace's school – one of Chloe's posho mates had tagged Chloe in some school fete thing a couple of months ago. Honestly, these rich women. No sense at all.

I called them up, doing the accent I'd heard on one of Chloe's videos, and told them I was sending my sister to collect Abigail later because I was stuck at work, then hung up before they could argue.

I was nervous when I turned up and asked for Gracie – or Abigail, as I was careful to say – but the teacher looked so harassed with the other kids she was pretty happy to hand her over to me when I said Chloe had got stuck at work and asked me to pick her up. I put on that accent, the one Sally's got and rolled my eyes when I said about Chloe's job as if she was the type of mum who's always late. Maybe she is. Maybe she doesn't even care about Grace. How can she? She didn't grow her in her literal body for nine months. She'll never have what Gracie and I do.

The moment I first saw Grace, my gorgeous little girl, I thought my heart was going to fall right out of my chest onto the pavement. Of course, I had to be calm – didn't want to scare her off or give that dumb teacher any reason to doubt me. Grace wasn't sure at first, said she wanted to wait for Mummy, but I said Mummy was busy and Sally and Nathan had told me what a good girl Abigail was for them when they looked after her, and wouldn't she like to be a good girl for me? She trusted me then; it only took knowing a few familiar names for her to believe I was someone safe.

I *am* someone safe. The only person in her life who is.

So now we're on a train with the five hundred quid I begged Layla to lend me. That's it, though. It's all I've got for now. I'm going to have to be careful – but I know there are always ways to make more money when you really need it. It's how I've managed my entire life.

We've been on the train for an hour and a half. We had to switch at Birmingham from Stratford, which is where we got on. I bought Grace a cap to match mine so people wouldn't easily see her face but there's been barely anyone on the train since we got on.

When we pull up to the next station though, there are hundreds of people waiting on the platform. I check my phone. It's rush hour. As the people start to pile into the carriage, I turn my body towards Grace so she's partially hidden. The gentle thud of her heart presses against my chest and I close my eyes, remembering the last time we were together like this.

It was the day of her birth, lying on the bathroom floor in my mum's old flat; I was off my tits on something Shane had given me – because they'd have given me drugs at the hospital anyway, so what was the difference in doing it myself? – but I still remember it. My warm, sticky little baby girl was lying on my chest, her breath so soft you'd almost think she wasn't breathing at all. She lay like that for what felt like hours but probably wasn't, then eventually I was alright enough to stand up with her. I cleaned her up like they'd do in the hospital, then we got into my bed together, a warm precious parcel. My little girl, safe by my side. I've never been so proud of anything as I was that day, looking at this tiny little life I'd created. Seeing her, so warm and safe, just the two of us, was what made me realise for sure that I had to leave Shane.

I sigh, thinking how stupid I was back then to ever believe things would be that easy. I properly thought Nathan and I would walk off into the sunset together: me, him and Grace. The perfect little family. Then he fucked everything up.

'What are you doing?' Gracie says. 'You're hurting me.'

'Sorry,' I say, releasing the grip I didn't realise I'd held so strong around her small shoulders and she pulls back, away from me.

She scowls at me and a flicker of annoyance curls in my stomach. But I crush it down because this little girl has no idea I'm actually saving her. Once she gets to know me, realises who I am and what I'm doing for her, she'll love me. I know she'll love me. I think of that perfect little baby lying on my chest, how she knew nothing and needed nothing in the whole bloody world apart from me. If everyone else would leave us alone, it could be like that again. I know it could.

The train pulls away but now there are people standing in the aisles and filling all the seats surrounding us. I feel watched, like a zoo animal.

In front of us, two women in smart clothes moan about their day at work and I listen into their conversation to try and take my mind off things but after a while they stop chatting as one of them pulls out her phone and starts scrolling instead.

'Have you seen this?' she says after a couple of minutes. She holds her phone up to the other woman.

'Shit,' the woman says. 'Honestly, though, why would the school just let a kid walk out with a stranger? Idiots.'

I flick a glance at Grace but she's too busy scowling at her own reflection in the window to realise what's happening. I lean forward in the seat to see the other woman's phone screen in the crack between the chairs.

Shit. Shit. Shit.

Grace's smiling face is plastered all over the news.

We need to get off this train.

Right bloody now.

At the next stop, I pull Grace's cap further down her head and pick her up like a baby. I expect her to struggle but she doesn't, she just lets herself be carried with her head buried in my shoulder. I keep my own head down as we shuffle through the busy train carriage, expecting someone to grab my shoulder and wrench Grace from my arms any second, but nothing happens and after a moment we're out on the platform. I don't put Grace down as we go through the barriers and luck is on my side because there are so many people at the station the attendants are holding the barriers open, so we don't even need a ticket.

I don't take a breath until we are out of the station and onto the street.

CHAPTER THIRTY-FOUR

Things start falling apart quite quickly after that. I thought Grace was happy to be carried but it turned out she was only still 'cause she was terrified and when I set her down on the pavement a couple of streets away from the train station, she starts bawling.

'I want my mummy,' she cries again and again and again.

'Please, please stop crying,' I beg her, looking around to check no one else is watching this. The street is empty, thank god, and I grab Grace's hand and lead her towards a park in the distance. She pulls away from me, so I pick her up again but this time she's stiff, all bony elbows and hard shins.

She's heavy, and sweat trickles down my forehead from the weight of her as my back screams in protest.

If I'd had the chance to know her, would it be easier to carry her now? Would I have developed the kind of strength a real mum has with big strong bulky arms? My muscles burn with every step and it's such a shit reminder of how fucking unfair this whole thing is. I should be able to carry my daughter. It shouldn't be this bloody hard.

When we reach the park, I basically drop her down onto the bench and stand bent over for a second to get my breath back, my chest burning. Grace sniffles and it hurts me, to see those tears pooling in her eyes that are so much like my own.

'I want to go home,' she says quietly.

I sit beside her on the bench and take her small hand in mine.

'Grace—'

'That's not my name!' she squeals, and I bite my lip.

'I'm sorry,' I say. 'But I need to tell you something really, really important. Are you ready to listen?'

Grace looks up at me and nods solemnly, her cheeks red and blotchy. I reach forward to wipe the tears from her cheeks, but she jolts back and I hold my hands up in surrender, pushing down the hurt in my stomach.

'Your name actually *is* Grace,' I say, and she shakes her head slowly. 'It is. I know it is because it's the name I gave you when you were in my tummy.' I pat my stomach softly. She watches and then looks back up at me. 'I'm your real mummy, Gracie.'

She doesn't say anything for a moment, studying me with those big brown eyes.

'Did you know you were adopted?' I ask.

She nods. 'Because my tummy mummy couldn't keep me safe.'

Tummy mummy? My face must show how cross hearing this makes me as Grace's eyebrows shoot up and she shifts away from me on the bench. *Tummy mummy?* What the fuck sort of shit have they been feeding her?

'That's not true, sweetie,' I say through gritted teeth. 'I tried everything I could to keep you safe but the big bad people came and stole you from me. Like a trick.'

Grace doesn't say anything again, but she looks scared and I hate that they've done this to her. To us.

'I wanted to keep you so badly, but the police came and took you away.'

'Why didn't you come back and get me?'

I swallow, trying to find the words. 'I've been trying, sweetie. But it's been really hard because the bad people were keeping me away.'

She thinks about this for a second as she plays with the hem of her jumper. Eventually, she says, 'Were you locked away in a tower?'

I smile and nod. 'Sort of, yes.'

'Like Rapunzel.'

'Yes, exactly like that. But I'm free now and I came to find you and be your mummy again.'

Am I really free? Sally knows all about me now. She's been to my flat; she knows how much I want Grace back. Is she ever going to let me go?

When I found out she was marrying Nathan, I honestly thought she'd set me up – that she'd created this whole email scenario as a ploy to keep me close, to spy on me for him. The casual way she threw in his name in her email about getting married, *everyone already knows me as 'Yatesy's Mrs'*, as if he was no one special. I felt like a total fucking idiot. How had I not known all this time that Sally was going to marry the man who ruined my whole fucking life?

But then again, I was the one who fell for a pig.

It was months after Shane's murder before I even worked out who Nathan really was. *Lee*. That's the name he used when he was palling up with Shane and getting too friendly with me. All that time, he'd been undercover. It was the newspaper article about Sally's award that tipped me off about who he really was – what I'd got myself caught up in.

I smelled a rat the night Shane had been killed, but I never really understand how Nathan had the power to force me into hiding. Then, looking into the local news for stories about my Gracie, I found the write-up about what a hero Sally was. I wasn't lying when I told Sally I took her name because it sounded like the name of someone good – someone I wanted to be. What would have happened if I hadn't made that stupid mistake? After I read the article, I wanted to see Sally's face and so I went looking on Warwickshire Police's website. But it wasn't Sally's face I found…

That's when I knew the real reason I was being paid to stay away. It all added up. Everything that happened that night. It was far bigger than one stupid man and it was the reason why I was never going to be able to go home again, never be Natalie Pierce. Never be Grace's mum.

'So can we go home now?'

Grace's hopeful voice breaks my thoughts. I look down at her, my throat raw with a swallowed scream because the truth is we can never go home.

A mum with a buggy walks past and throws me a consolatory smile, as if I'm just like her. A stressed mum dealing with their whiny child. I pull my shoulders back, letting myself pretend for a moment this is true, that we'll cross the park and walk up the steps of our posh house with a little gate and a dog in the doorway.

'Can we, please? Can we go home?' Gracie asks again, the sadness of her voice bringing me back to earth with a crash. There is no posh house, no cute gate, no yappy little dog. I rub my face and take a deep breath, trying to keep the panic out of my voice when I reply.

'We're going to find a new home, just you and me.'

Her voice breaks and her little face screws up into a panicked frown. 'But what about Mummy and Daddy?'

I shake my head. 'I just told you Grace, I'm your real mummy. So you don't need your old pretend mummy and daddy, do you?'

Tears fill Grace's eyes again and I'm angry she isn't more excited that I've finally found her, that it's going to be the two of us like it was always meant to be. But then she is only five. She'll come around. The longer she spends with me, the more she'll realise she's my little girl, not those people's.

'What about Auntie Sally and Uncle Nathan's wedding? They need me to be a flower girl. It's on the weekend,' she says so seriously that it undoes me for a second. Tears fill her eyes and I know I have to calm her down.

'Why don't you tell me all about the wedding,' I say and for a moment it works. She's so distracted giving me every detail about the wedding that she walks hand-in-hand with me through the park.

'Don't worry, we'll definitely go to the wedding,' I tell her when she's done. 'But for now, you're going to have to be a really good girl for Mummy, because the people who took you away want you back and so they've sent out lots of spies to look for you.'

Her eyes grow wider, and she flicks her head to look around the empty park. 'Where?' she whispers.

'Everywhere,' I say. 'So we're going to play a hiding game, okay? Just you and me. We're going to hide away until the big bad spies have disappeared. Can you do that for me? Be really, really good at hiding, like a big brave girl, so the bad guys don't catch you and take you away again. Okay?'

Grace nods and says in the smallest, saddest voice I've ever heard, 'Okay, I'll be good. Promise.'

I bring her in for a hug and, while she doesn't wrap her arms around me like I want her to, she also doesn't stop me from holding her and I think that's pretty much all I can ask for at the moment. There's a second where everything is perfect, just me and my girl on a sunny evening in the park.

But then, I see a familiar face.

Quickly, I form another plan. Perhaps even more stupid than the last.

'Gracie…' I whisper instructions in her ear.

She listens, then nods at me, smiling and I have to hope she's understood.

'Okay and now, when I count to three, I need you to run. Do you understand? We're going to run across the park to that gate over there, do you see?' I'm careful not to point or lift my head, so it's not obvious I'm giving her instructions. She nods as her heart thuds against my chest. 'Okay, ready?' I say. 'Three, two, one…'

She's up before I manage it myself, a flash of red running across the park and I think we're going to make it, I think we'll get away but then everything comes shattering down as I'm tackled to the floor – I watch my beautiful girl run ahead without me as I lie face down in the grass.

Fuck.

CHAPTER THIRTY-FIVE

SALLY

We stand on the shitty street outside the tattoo parlour looking at each other uselessly.

'What now?' I ask.

Nathan shakes his head. 'We're going to have to tell someone about you and Natalie. We can't risk letting something happen to Abigail just to save our own skins,' he says and I nod. 'It might not be that bad...'

I laugh. 'I've helped a known fugitive hide out, hit a man with my car, then withheld information about a child abduction case. I think it's going to be fucking bad, Nath.'

His jaw twitches as my words settle.

His phone rings and he sighs, then answers.

'Hi, Mum,' he says, turning away from me. 'What?' He turns back after a few seconds of listening to whatever Amanda is telling him, a smile forming on his tired face. 'Where?'

I mouth 'what?' and tug on his sleeve but he waves his hand at me to silence my questions then says the words I've been desperate to hear.

'They've found her. Abigail. She's safe.'

The street around us falls into silence – no longer do I hear the angry squawk of the seagulls or smell the stale stench of cigarettes and piss from the nearby alleyway –

relief washes over me as I close my eyes and take a deep, salty-aired breath. Abigail is safe. That's all that matters.

I open my eyes and things come into focus again, along with a sharp dose of reality.

If Abigail is safe, does that mean they've found Natalie? And if they have, is she about to tell them everything I've done?

Nathan finishes his phone call and pulls me into a hug, a tang of sweat coming off his body as he presses my face into his musty t-shirt.

'Nath,' I say, pulling back. 'What about—'

'Natalie is gone,' he says, a smile breaking out over his face. 'Mum said the police found Abigail in a park near the station in Bristol, on her own. She said Natalie had dropped her there and run.'

I frown. 'I don't understand. Why would Natalie go to all that effort to take Abigail only to abandon her again?'

Nathan shrugs, unbothered by this turn of events. 'I guess she realised she'd never get away with it – especially if Abigail started playing her up. You know what she can be like when she's away from Chloe. Anyway, who cares why? The point is, Abigail is safe and Natalie has disappeared again. Hopefully this time, she'll stay gone.'

I sigh, picking at the skin around my nails.

'Sal,' Nath says, taking my hands in his and running his fingers over where I've picked until it's bled. 'This is good news. If Natalie stays gone, nothing ever has to come out about you. This is the best possible outcome; Abigail found, and Natalie gone.'

'I just don't understand why she'd run away again,' I say.

Nathan drops my hands, his face clouding. 'Why try and understand a person like that, Sal? She's probably got off her head on something and then thought kidnapping Abigail would be a good idea. I doubt she thought much further ahead than that.'

'You say it like you know her,' I say, and he takes a step back.

'Sal,' he says, his face softening. 'Can you not just be happy for one minute that Abigail is safe?'

I study his face, so unsure of everything now. Why doesn't this feel good? Why am I still worried about Natalie, despite everything she's done? Why am I so desperate to believe that it's not so easy for someone to abandon their child?

'Let's go home,' Nath says.

But I shake my head. 'I need to know what happened between the two of you.'

He nods. 'Okay,' he says. 'But can we at least get away from this shithole first?' He smiles, the dimples in his cheeks revealing themselves and I nod, take his hand, and follow him into the car.

—

Despite planning to grill Nathan the whole way home, I wake up groggy in the passenger seat as we turn into Stratford town centre.

'Hello, darling,' Nath says, glancing over at me with a soft smile.

The night has drawn in and Stratford is lit by the soft glow of the streetlamps; a couple of people dressed up on their way back from a night out at the theatre walk down the street, laughing. I clack my tongue against the roof of my mouth. It's dry, like I've been sleeping with my mouth wide open.

'We're home,' I say, trying to get my bearings.

'Almost.'

'I slept the whole way?'

He looks over again and smiles. 'Not surprising. It's been a pretty stressful day.'

And it's not over yet.

Nath pulls the car into our designated spot and comes around my side to open the door, like we're in an old film and he's my suitor. He guides me out of the car like a small child, sleepy after a long journey, then leads me into the house.

Inside, I follow him upstairs to our bedroom and watch as he pulls back the covers to our bed. It would be so easy to simply shrug off my clothes and climb in, go back to sleep, and wake up tomorrow as if nothing bad has happened. But I can't. I can't keep hiding from this.

'Nath,' I say, as he turns to leave the room. He stops. 'Please. We have to talk about this.' His shoulders slump and for a second I think he's going to keep walking, to leave the room and this conversation but then he turns around, a look of resignation on his face.

'I'm knackered. Can't it wait until tomorrow?'

I shake my head, knowing if we don't tackle it now, I might never get the nerve up to face it again.

He sighs, then walks over to the bed and sits awkwardly on the edge.

'What do you want to know?' he asks, his eyes on me.

I fiddle with the pom-pom on the edge of our pillows. Nath chose these; we never fit into the stereotype of me being the one who chooses all the decoration and him having no clue. I've never had a good eye for things in the house, but Nath has always known what goes with what.

'The truth,' I eventually say, looking up at him. He nods, then starts to speak.

'What I said earlier today in the car was true,' he starts. 'I was undercover trying to get the head of the drug and people trafficking operation that Shane was part of taken down.' He pauses. 'And I did meet Natalie then.'

'Did you have a relationship with her?'

'No,' he says firmly, and I search his face for the tells of a lie but I'm so tired and strung out from the day it's like none of my senses are working. 'But she...' He sighs, rubs his face like he's embarrassed. 'She liked me. A lot.'

'What do you mean?'

'It got a bit out of hand,' he says, looking down at the duvet cover and brushing his hand against it to flatten the creases. 'At first, I thought she was okay. I felt sorry for her, protective, I guess. She seemed so young to be caught up in all that—'

'How old was she when you met?'

'Nineteen,' he says, without a beat, and I wonder if this is a number he's recited to himself before – because it's old enough, isn't it? Technically. 'Sal,' he says, his eyes blurring with hurt. 'Please stop looking at me like that.'

'I'm not looking at you like anything.'

'You are,' he says. 'You're looking at me like I've done something wrong. And I did, but not in the way you think.'

My heart skips at the admission. 'What did you do?'

He sighs again and avoids my gaze. 'Like I said, things got out of hand. Natalie had this stupid crush on me – it was one sided, I swear – but Shane soon picked up on it and, well, it just screwed everything up.' He huffs, red spots marking his cheeks. 'It was fucking embarrassing, Sal. It was the first undercover job I'd been put on – it was all

I'd ever wanted, to go undercover, do things that actually mattered instead of speeding offences and neighbourhood talks,' he says and I nod because I remember him talking about this when we were young, how he was desperate to make a difference – to not just be a beat cop, doing the same every day. 'But instead, I end up pissing the whole operation up because some teenager got a mental crush on me and wouldn't stop trying it on.'

I narrow my eyes, trying to imagine a young Natalie relentlessly pursuing Nath. 'Why would that mess the operation up though?'

'Because Shane stopped trusting me. As soon as he saw Natalie looking at me the way she did, he didn't want me anywhere near her. That meant I got shut out, completely.'

I nod, understanding. 'So they pulled you off the op?' I ask and he nods.

'I'd barely got any information at all; I'd only just started. The lads at HQ found it hilarious too, called me The Heartbreak Kid for months afterwards.' His cheeks shine with embarrassment and I want to reach out and comfort him because I remember what Nath was like back then, despite being in his late twenties, he had not yet grown into his features – he was still handsome, still charming, but compared to the bulky blokes on the force he looked like a baby. It was probably why they wanted him undercover in the first place, because he could pass for someone much younger without raising any suspicion.

'I don't understand,' I say. 'What's so bad about that? Why wouldn't you just tell me?'

'It's not that,' he says. 'Getting taken off the operation was embarrassing, but not the end of the world. It was what happened after.'

We stop as the sound of a group of lads walking under the window drifts into the room, their voices loud and lairy from the pub – a world away from the night we're having.

Nath continues. 'Look, I know what you're thinking – Natalie was nineteen and vulnerable, with no one to protect her and she was under the control of a violent thug. And that is true. But Natalie was also... She was a nasty piece of work,' Nath says, his mouth chewing over the words like they leave a bad taste. 'I know I shouldn't say that. But she was. She wasn't just Shane's girl; she was involved in everything. She was the one who would bring him the girls. She'd be sent out to find the youngest ones she could and convince them to sell themselves, for him.'

I shake my head. 'Come on, Nath. You know that's how it always works, girls like her get forced into it and then manipulated into forcing others—'

He cuts me off. 'No, Sal. I'm sorry, but there was something about Natalie. Yes, she was probably forced into it at the start but by the time I met her, it was like she enjoyed it.'

I search his face, trying to piece together this information with the woman I've met, the woman I've spent years emailing. I've convinced myself she's a poor innocent victim in all of this – believed her when she said she was set up – but what if it's all been a lie?

'What does this have to do with what happened that night?' I ask. 'She said she was set up over Shane's death. Is that true?'

Nath shakes his head, sadly. 'I was there that night,' he says. 'It was months after I'd been kicked off the operation, but Natalie and I had... Well, we'd kept in contact.'

I narrow my eyes at him.

'I know it was stupid, but like I said, she was obsessed with me, and I thought… Well, honestly I thought I might be able to use that to my advantage. I thought if I kept in contact with her for long enough, she'd eventually tell me something that would uncover the whole trafficking ring.'

'You used her to get back into undercover work,' I state.

'I know it was wrong. But I was young, and stupid. Meanwhile, she was getting more and more obsessed with me – wouldn't stop talking about us running away together.' He frowns and I think about what Natalie told me a few days ago, about how she loved Nath. Maybe she did. 'I had no idea she'd had a baby,' he says. 'She must have been pregnant when I was undercover, but I never noticed—'

'You didn't notice a pregnant stomach?' I ask sceptically.

He shrugs. 'She was tiny, I don't know. She hid it well, I guess.'

I gesture for him to go on.

'But then this one night she gets all desperate, asking me to meet. We hadn't seen each other for months – not since I'd been taken off the op – we'd only texted, I thought it was safer that way. But that night, she said she had something big to tell me – something that was worth my while.'

'You thought you were going to get the information you needed,' I say, and he nods.

'So I turned up, and knew straight away it was a mistake,' he says, eyes down now, not able to look at me. 'She was feral, on something, I think – kept talking about how she was finally going to be rid of Shane once and for all, laughing like she had the most amazing plan. We argued; I told her I didn't want anything to do with it, then Shane showed up.'

'You were there the night Shane died,' I say, feeling sick.

He nods. 'But I left,' he says quickly. 'Before anything happened, I left.'

There's silence between us while I take in what he's saying.

'Did anyone know you were there?' I ask and, after some hesitation, he nods.

'Who?'

He sighs. 'I heard the call come in, about Shane, once it all went down and he was killed. I was on shift. I knew I'd fucked up. I was terrified they were going to get Natalie's phone and realise we'd been talking.'

'What did you do?'

'I needed help, and there was only one person I could think of who might know what to do…'

The silence sits between us uncomfortably as it dawns on me who Nath would go to in a crisis; a man he'd looked up to his whole life.

'My dad,' I say, sadly. 'You asked my dad for help.'

He nods, looking up and reaching out to me, but I pull my hands away. 'He made some calls. I don't know exactly how he did it, but he got Natalie's phone before anyone else could. We got rid of it. No one ever knew I was there that night, or that I'd arranged to meet her. If they had…'

'You'd have lost your job and potentially been involved in a murder case,' I state simply.

'I know,' he says. 'It's bad. It's really bad. But look, that's all that happened. I fucked up. I made some stupid decisions, and I hid them from you — but your dad told me to. He said we could never tell anyone about this.'

'That's why he said it...' I say more to myself than Nathan.

'Said what?'

I sigh. 'Before he died, my dad told me not to marry you. I never understood why,' I say. 'But now I do...'

Nath's face cracks then, and the look of heartbreak is so painful I have to look away. He loved my dad; hearing this will be hard for him.

'I thought he understood,' Nath says, like a lost little boy but then he shakes himself out of it. 'Look, Sal, the point is – Natalie is dangerous. I think she knew you were with me and that's why she took your name and found a way to contact you.'

'You think this is about you?' I ask. 'You said it was about Abigail earlier.'

'I didn't want to freak you out, and I'm sure a lot of her reasons for keeping in contact with you were to do with Abigail, but you must admit it seems a bit too much of a coincidence she took your name – the woman I'm about to marry – out of all the names she could choose, isn't it? I don't think she ever got over the idea that me and her were going to run away together. I think she's manipulated you, like she manipulated all those young girls back then.'

Maybe he's right. Maybe I am as gullible as they come, believing in the idea of Natalie the victim, Natalie the helpless. Instead of seeing her for who the rest of the world sees – a murderer and villain.

But where does that leave us? What are Nath and I, if not villains too? We've both done things we shouldn't have – things that could land us in prison – and yet we're here, we're free and Natalie is the one in hiding.

It's too much to think about, too much for one night.

For now, all I can do is sleep.

CHAPTER THIRTY-SIX

It's been a couple of days since Natalie abducted Abigail and no one has heard from or had any sight of her since. Nathan and I have been cautious around each other, moving around the house almost like strangers. He's polite and delicate with me, offering to make endless cups of tea and doing the washing up the moment we've finished dinner instead of leaving the dirty dishes in the sink for hours on end until one of us relents and begrudgingly cleans them.

Today, he's on shift while I'm on a rest day and I haven't yet managed to get myself out of bed. It seems our escapade down to Weston has gone unnoticed, and so we've faced no consequences for our actions. At least, not professionally.

The wedding is two days away and yet we've barely discussed it. I get the sense Nath is worried to bring it up in case I tell him I'm calling the whole thing off. I can't say I haven't thought about it. Can I marry a man who lied to me for so long? He says it's not lying; he just never told me. My dad's warning that I shouldn't marry Nathan repeats in my head. Dad knew, all this time. He knew what Nathan had done. The ridiculous situation he'd got himself into.

I hate Natalie. Every time her face appears in my head, a coil of rage unfurls. I was so stupid for believing in her, for believing she had been wronged by the system and thinking that deep down all she ever wanted was to be a good mum. She never cared about Abigail – if she did, she wouldn't have left her that night and then abducted and abandoned her all over again. *When people show you who they are, believe them the first time.* That's what my dad would always quote when I started mentioning an interest in finding my own mum. *She left us, Sal,* he'd say, the hurt blooming across his face. *I wish it wasn't true, but it is – she didn't want to be your mum.* I used to rally against it at the start, believing that people were mostly good, even when they made mistakes, but the longer my mum stayed gone, the more I knew my dad was right. If my mum wanted me, she would have stayed. And if her leaving was a mistake – a blip, a stupid decision – she would have come back for me. But she never did. She proved herself to be exactly the person my dad had told me she was. Selfish. Callous. Unfit to be a mother.

I should have remembered that women like that don't change. If I had, perhaps I wouldn't have been taken in by Natalie's woe-is-me act. All I did in trying to save her was put Abigail at risk.

I sigh, rubbing at my tired eyes. How can I be angry at Nathan for his past mistakes when I've made so many of my own?

The doorbell rings and I consider ignoring it. I'm not dressed despite it being eleven a.m. and the idea of letting someone know this – even the postman – is less than desirable. But then the ring is followed by three quick but firm knocks and I know Amanda is on my doorstep.

Grabbing a hoody from the back of the door, I pull it over my head as I pad down the stairs shouting, 'Coming!'

'Morning,' Amanda chirps when I open the door. She holds up a takeaway coffee cup with a smile.

'Oh, thank you,' I say, taking the cup. 'I didn't know you were coming over?' I step back to let her in and she walks through the hallway into the kitchen where she takes a seat at the table. I follow her in and sit down, the warm coffee in my hands.

'I had to pop into town to pick up something for Abigail,' she says. 'Thought I'd stop in and see you while I was here. Nath said you were on a rest day.'

'How's Abigail doing?' I ask, taking a sip of my too-hot coffee so that I don't have to meet Amanda's eye.

'Oh you know,' she says breezily. 'Chloe said she hasn't mentioned it much since she got home. I think it's Chloe who's suffering more, to be honest.'

'Of course,' I say. 'Must have been terrifying for her.'

'Mmm,' Amanda says. 'But still, Abigail's back safe now and it won't happen again. The school have been read the riot act, that's for sure! Imagine letting a stranger take her off like that. Absolute imbeciles.'

'Has she gone back to school?'

Amanda shakes her head. 'Chloe won't let the poor girl out of her sight! Still, I can understand that. She's terrified it'll happen again, but I've told her it won't.'

'How can you be so sure?'

Amanda looks at me with a raised eyebrow. 'That woman won't come back,' she says, utterly certain, and I wonder how it must feel to be that confident about anything. 'I've no idea what she was playing at, taking the poor girl like that and then abandoning her again. It's despicable.' Amanda speaks with such hatred I'm taken aback, but then I remind myself that Amanda has loved Abigail like her own grandchild for her whole life.

'I just don't understand why she did it,' I say.

'She probably had an idea in her head that Abigail would run off into the sunset with her and when faced with the reality of having a child – even for a few hours – realised it wasn't quite what she had in mind so decided to run off. Sadly, some women are not meant to be mothers. They have no understanding of what it means to be one! You know this, of course.' She reaches out and takes my hand and I look away.

Is Natalie so different from my own mother? Like her, she ran away the moment it got hard – the moment *I* got hard. Is that why I wanted to believe Natalie was good, that she hadn't chosen to leave Abigail but had been forced?

God. Deep down, am I still clinging on to the idea that my own mum had no choice but to leave me? If she'd had a choice, she'd have come back? Is that what this whole thing has been about?

'So, the big day fast approaches!' Amanda says, breaking me out of my existential crisis. 'Let me show you Abigail in her little dress, she's so adorable.' Amanda reaches into her bag for her phone and proudly pulls up a photo that, admittedly, is adorable and I force a smile.

If I don't go through with the wedding, there will be so many people left disappointed. I try and imagine telling Amanda I'm having second thoughts, but the idea is ridiculous. Unlike my own mother, Amanda has always been there for me. Without my dad, she and Nath are the only family I have left. How can I possibly imagine a life without them?

'It will be good for all of us, the wedding,' Amanda carries on. 'A reminder of what life is supposed to be like – no abductions and emergency police phone calls. Just us, all together. Celebrating what really matters.' She nods decisively, still smiling at the photo on her phone.

She's right.

This wedding is exactly what we all need – to move forward and forget about Natalie and her sordid world of chaos and crime.

I take out my phone and message Nathan.

> Let's not talk about what happened any more. It's in the past. I know you're a good person – you're my good person, and I can't wait to marry you this weekend. I love you.

I press send before I can talk myself out of it and look up at Amanda smiling at me.

'Messaging Nath?' she asks, and I nod. 'Look, why don't you go up and get dressed and we'll go out and get some lunch? It's a lovely day, far too nice to be cooped up inside.'

'Yes, okay,' I say.

'You can help me make the final decision on my mother-of-the-groom shoes, too,' she says, laughing. 'I've narrowed it down to two pairs from Next but you know what I'm like, I'm not built for heels.' She frowns, gesturing at her admittedly sturdy calves and I laugh. 'I've always preferred shoes I can run in, if the need arises.'

'Well,' I say, my mood lightening. 'There will certainly be no need for you to be running anywhere at the wedding, so I think in this case, heels are appropriate.'

PART FIVE

CHAPTER THIRTY-SEVEN

SALLY

I look at my reflection in the mirror. I've never thought this about myself before but today, I look beautiful. It's amazing what a professional make-up artist and hairdresser can do. I tilt my face to one side, then the other, marvelling at the way she's made my cheekbones look sculpted.

'Don't do that!' the hairdresser calls from behind as she watches me tuck my hair behind my ear.

'Sorry,' I say. 'Habit.'

She shakes her head and waves goodbye as she picks up her enormous black bag of tricks and leaves the room.

Now it's just the two of us and Amanda appears behind me in the mirror, her hands on my shoulders.

'Sally,' she says, her eyes glistening. 'You look…' She shakes her head gently, like she doesn't have the words. 'Your dad would be so proud.'

I swallow before replying, the words catching in my throat. 'Do you really think so?'

'Of course, darling.' She squeezes my shoulders and smiles tightly. 'I'm sorry he can't be here, but it's a real honour to be able to walk you down the aisle, you know.'

'You're the only person I would ever want to do it, other than Dad,' I tell her, because it's true. It might be unconventional to ask your mother-in-law-to-be to walk you down the aisle but it makes total sense for us. I don't think I'd be here today, or be the person I am, without her.

Amanda looks like she might cry, which I find weirdly unnerving; Amanda isn't a crier. Amanda is strong and stable and reassuring – everything I find myself unable to be lately. I knew today was going to be hard without Dad, but it's worse than I imagined. I'm trying so hard to push the thoughts away – the idea of him standing here with me like I always thought he would, at the perfect wedding. And if I'm honest, it's more than his absence upsetting me. It's the thought of him being here, and being unhappy about it.

You shouldn't marry him.

The sound of my dad's voice in my ear is louder than ever today – though I've chosen to put whatever Nathan got up to with Natalie behind me. I've chosen to not look too closely at what my dad might have got himself involved in, either. What's the point?

'Come on, you don't want to leave Nath waiting at the altar,' Amanda says, opening the hotel room door. 'He'll be terrified, standing up there alone.'

She's right. I can't leave Nath waiting any longer, and so I silence my dad's voice in my head and stand up – taking one last look at myself in the mirror, the perfect blushing bride, and go.

–

'We did it. We actually did it!' Nath turns to me with the biggest grin on his face and I grin back.

'We're married!' I say, like something from a half-written film script.

He pulls me in for a kiss – the second of our married life, and we laugh.

It's just the two of us now, in the back of an old classic car he rented for us, each holding a glass of champagne.

'I love you, Mrs Yates,' he says.

'I love you too, Mr Yates,' I say back, with only a hint of sadness that Sally Jones is no more, and this thought brings a sharp reminder of the woman that was once the other Sally Jones, out there somewhere still.

Our driver gets in and the car purrs into life. We drive the five minutes to the venue in silence, my hand clasped in Nathan's as he stares out of the window with a Cheshire cat smile.

We've done it. We're actually married.

Dad is on my mind, of course, but the dad he was before the dementia started taking over. Before he had those weird thoughts about Nathan. I think of the dad who would watch Nath and I playing as kids, who'd scoop us up in one arm each and throw us over his shoulders like a fireman as we'd scream. That dad would be so proud of me today; he'd be so happy to see me with Nathan. To watch me become his wife. That's what I'm going to remember throughout the rest of the wedding, not the doubts I had sitting in my dressing room, not the absolute insanity of the last two weeks with Natalie. I'm moving on now – this is our new life. Nothing that happened before matters.

'Mr and Mrs Yates,' the driver says as he pulls up to the venue. He looks over his shoulder and smiles before getting out of the car to open Nathan's door. Nath walks around the side and opens mine for me.

'My wife,' he says, holding out his hand like an old romantic.

As I step out of the car, our guests are lined up outside the venue, smiling and clapping. Part of me wants to get straight back into the car, for it to be just Nath and me again, alone and quiet without all of these eyes staring at me, but Nath beams and takes my hand in his, fingers interlaced, as we walk towards the crowd.

'Congratulations!' The word is said so many times as we walk up the steps of the venue past our guests that it becomes one big blur of a word. I smile and say *thank you, thank you, thank you*, like none of this is quite real.

Nath walks me through the guests, down the corridor and into our drinks reception. I look around the room, amazed at how beautiful the space has become. When we came to look around it, it was fairly plain. Amanda must have put so much work into making the venue look and feel so special. I touch the flowers I've had nothing to do with picking, lifting them to my face to smell the sweet scent of the peonies. My favourite.

'Everything looks so beautiful,' I say to Nathan.

'It really does,' he says but he's less enamoured by our surroundings than me, instead busy spotting friends who pile through the doors and soon he lets go of my hand and greets them like a good host.

—

We decided to forego normal speech traditions, as I have no father alive to speak on my behalf and so after Nath's best man – Andrew, an inspector he met on his first year in the force – gives a long and pretty funny speech detailing some details of Nath's stag-do I'd rather not have heard, Amanda stands up for her turn. The room instantly quietens, like the position she once held at the police still applies here, no matter what.

'Thank you, Andrew, that was...' Amanda pauses, and everyone laughs slightly nervously. She raises her eyebrows but says nothing as the laughter continues – more joyous now. 'I'm delighted to be standing here with you all today and would like to say thank you on behalf of both my darling son, Nathan, and my beautiful new daughter, Sally.' Amanda beams at me and I blush under her gaze while everyone cheers, all the nervousness seeping from the room. 'As you know, these two have been destined for marriage since they were this high,' she says, her hand down below the table. 'I can't believe the little girl who used to put a pillowcase on her head and the little boy who proposed to her with a Haribo ring are finally married...'

I sit back in my seat and listen to Amanda's words, a warm sense of calm flowing through me – finally. Nath clutches my hand and turns to smile at me throughout, with each anecdote about our lives together further securing the idea that today was inevitable. Written in stone before we were old enough to even understand it.

Nathan's speech is similarly perfect – he tells the room how lucky he is to have found me, how much he wishes my dad were here to see it and this time I don't flinch at the reminder of my dad but instead play those scenes in my head again of Dad happy, Dad smiling. Nath finishes by telling the room how grateful he is for everyone who has come to celebrate our love. The whole room is thick with emotion; it's a little overwhelming and by the time the speeches are done, I excuse myself to slip out into the reception area to gather my thoughts for a minute.

–

I'm sitting on the plush dark green velvet sofa outside the main room when there's a tickle against the back of my leg and then I hear a little giggle. Looking down, I spot a small hand under the table next to me and smile.

'Abigail?' I say, leaning down to look under the white linen tablecloth.

She is sat on the floor, her pink princess dress splayed either side of her as she puts her hands to her lips.

'Shhh,' she says. 'I'm hiding.'

I nod. 'I was hiding too, but you tickled me.'

She climbs out from under the table and plonks herself beside me.

'You look like a princess,' she says in a whisper.

'You do too.'

She nods sagely, like this is undoubtedly true.

'Does your mum know you're out here hiding?' I ask, surprised Chloe has let Abigail out of her sight for more than thirty seconds, given what happened with Natalie a few days ago.

'Which mummy?' Abigail asks in a sing-song voice, and I stop, unsure how to reply.

'What do you mean?' I say.

Abigail crinkles her nose, uncertain now. She looks down, playing with the tulle of her dress.

'I've got a secret,' she says, so quietly I have to bend my neck to hear.

'Do you want to tell me?' I ask.

She looks up at me and frowns, then shakes her head.

'Okay,' I say. 'You don't have to tell me anything, we can sit here and hide together.'

'Are you waiting for your mummy too?' she asks, and I tell her no, my mummy is gone.

'Where has she gone?'

I shrug. 'I don't know,' I tell her. 'She went away when I was a little girl like you.'

'And she never came back?'

I shake my head and Abigail thinks about this.

'My mummy is coming back,' she tells me, and I frown.

'Your mummy is in there,' I say, pointing to the reception room I've just escaped from.

Abigail shakes her head, sure of herself now.

'My other mummy,' she says, her big eyes looking up at me and I freeze.

'What do you mean?' I ask but Abigail doesn't say anything, just smiles. 'When did your mummy say she was coming back?' I ask.

'When we went on our adventure,' Abigail says. 'We made a special plan. Before Manda came and got me.'

'Okay...' I say, wondering what this means, but then Abigail continues.

'Manda shouted at my other mummy and then she said, "YOU STAY WAY OR THAT BE THE END OF YOU!",' Abigail imitates Amanda's voice and wags her finger in the air, her little cheeks puffing out and turning red with the effort of her – bad – imitation. She looks up at me seriously and adds, in her own voice, 'And Mummy told Manda she would stay away. She promised, but I know a secret and didn't tell Manda.'

'What secret do you know?' I ask and Abigail giggles.

'Other Mummy is coming today as a surprise, even though she told Manda she would stay away.'

'When did your mummy and Amanda talk?' I ask.

'When Manda come and got me! I told you!'

I shake my head. 'No sweetie, I think you're confused. It was police officers who came and found you, not

Amanda, your other mummy had already left. She left you alone, remember?'

Abigail shakes her head, her little round cheeks going red with frustration. 'No. I was with Other Mummy and she promised we'd go to the wedding together but then Manda came and made Mummy go away again. It's a game. Like hide and seek. Other Mummy made me promise I'd wait for her here today.'

I open my mouth, but words don't come out.

Amanda said Abigail was alone when the police found her – that Natalie was long gone. She certainly didn't say she was also there. Is Abigail confused about what happened and making this up? Or was Amanda lying?

'But you've got to keep it a secret,' Abigail says now, pulling at my arm.

'Where did your other mummy say to wait for her?' I ask.

'Here,' Abigail says, smiling. 'That's why I'm hiding!' She giggles like this is one big game – but it's not. If she's telling the truth, then Natalie is here.

Shit.

CHAPTER THIRTY-EIGHT

I need to find Nath. He'll know what to do.

'Abigail, sweetie, can you come with me?' I say holding my hand out to her.

She shakes her head, giggling.

'Come on, sweetie. We need to go find Nathan.'

'No!'

This time, her little face rearranges itself into an angry frown as she shouts. I've seen Abigail like this a few times before with Chloe, she goes from the angelic little girl we all know and love to the most stubborn dead weight you can imagine, and nothing will convince her to give in. She jumps off the sofa and hides back under the table.

I plead with her for a few more minutes, lifting up the tablecloth to see her, but the more desperate I become, the surer Abigail is about staying. I crawl under the table and try to drag her by her arms, which only results in a scream so piercing I crack my head on the table as I stand up to get away. One thing is for sure – there's no way I'm getting Abigail out on my own.

'Okay,' I say to her, sweat smudging my perfectly applied bridal make-up. 'It's okay, I'm sorry for dragging you.'

She looks up at me, her big eyes filled with tears – scared of the mad princess I've become in front of her.

'Can you make me a promise, Abi?'

She looks at me unsure, then nods slowly as she sniffs a bubble of snot that's formed on the end of her nose.

'You must stay here for me, okay? Stay hidden under this table and don't come out until I come back with Nathan. Okay?'

She nods again, solemnly this time as if I have come and ruined her game completely.

I take her at her word, because I have no other choice, and rush back into the main room. The atmosphere is like another world and it hits me: this is my wedding day. The day I married my childhood sweetheart, and instead of enjoying it I'm panic-stricken looking for my new husband, not to drag him for a dance to celebrate our marriage, but to warn him a mad woman is on her way to ruin everything. Again.

'Sally!' Cass, Nath's aunt, stops me with a huge smile on her face and a glass of prosecco in hand. 'Congratulations, darling. You look fabulous,' she says, leaning in for a kiss on one cheek.

'Thank you,' I say. 'Have you seen Nathan?'

Cass looks around then says no. 'Let's get you a drink, my darling. A bride should never be without a glass of bubbles!' She leads me away, towards the bar, but I shrug her off.

'I need to find Nathan,' I say, still scanning the room for him.

'Gosh, you've got your whole life together – you don't need to panic!' she says, laughing but then gives me a look like I'm being rude, which I am. But I've no time for this; I turn around and ignore her, walking in the opposite direction, her eyes and those of our other guests on my back.

I pass Chloe and Matt, who are laughing along at a story of Amanda's, oblivious to the danger their daughter is in. I almost stop them, tell them what's going on – but how can I explain it? I need to find Nathan. He will know what to do.

I walk through the whole room, ignoring every guest who tells me I look beautiful or wishes me and Nathan congratulations. As I'm about to give up and go back out to Abigail and forcibly drag her from beneath the table, screams be damned, I see him.

'Nathan!' I shout across the room, and he spins around, concern on his face at the desperation in my tone. He pats the arm of the man he's with – one of the officers I vaguely recognise from work – and makes his way towards me.

'What's wrong?'

I shake my head and drag his sleeve towards the door; he follows without question.

'It's Natalie,' I say when we get into the silent lobby.

'What?'

'She's coming here,' I say breathlessly, rushing towards the table where Abigail is hidden.

'Here? What do you mean? Sal—' Nath grabs me by the arm and spins me towards him. 'Stop. What are you talking about?'

I shrug him off. 'She's coming to take Abigail.'

He frowns and takes a step back from me as I turn back around and lift the tablecloth.

'Abigail, you can come out now,' I say. But it's too late. Abigail is gone.

CHAPTER THIRTY-NINE

'She's gone, Nath… she's gone. She was right here, I told her to stay right here.' I'm babbling now, Nathan looking at me like a mad woman as I crawl underneath the table on my hands and knees, my beautiful white dress dragging along the rough carpeted floor as if Abigail might be found if I can just get closer to where she was when I last saw her.

'Sal, stop, please,' Nath says, his hand on my shoulder, guiding me up.

I crawl back out.

'I shouldn't have left her,' I say. 'I'm so stupid. Why would I leave her?'

'Sal, please. Tell me what's going on. You said Natalie was here, that she was coming for Abi. Did you see her?'

I shake my head. 'Abigail told me. I found her out here, hiding. She said Natalie had told her to wait for her, here.'

Nath frowns, scratches the back of his neck as he thinks. 'Are you sure Abi hasn't just gone back inside?' he asks, nodding towards the main door hopefully.

'I don't know. Maybe,' I say.

'I'll go and see. Wait here,' he commands.

'Nath, don't say anything. Please, not to your mum or to Chloe—'

'I won't,' he says. 'Not yet,' he adds ominously.

He's only gone for a minute, but I count the seconds as if they each last days. He'll find her, he's right. I'm sure she's gone back inside, probably to tell her mum how insane Aunty Sally has gone. It'll be fine. Maybe I misheard what Abi was saying – or got it wrong. Natalie isn't coming. Why would she come here? She wouldn't risk it. Plus, what Abigail said doesn't make any sense. She said Amanda had told Natalie to stay away – but Natalie abandoned Abigail in the park before the police found her. Amanda wasn't even there. They didn't talk. That's what Amanda told us. If Amanda had caught Natalie red-handed, she would have made sure she didn't escape – Amanda would have held her there until the police came to make an arrest. She wouldn't have let her get away; she wouldn't have let this happen.

I'm so convinced by my own reassurances that Abigail has simply gone back to the party that when Nath comes back through the doors, relief rushes through me, but then he shakes his head and my legs give way.

-

'She can't have gone far,' Nath says once he's sat me down on the big, plush green velvet chairs in the lobby. 'I'll call it in.'

'No,' I say. 'You can't.'

'What do you mean? I have to!'

'Abigail said something… before I left to find you, she said something about your mum.'

Nath's eyebrows draw together in a frown. 'What?'

'She said your mum had spoken to Natalie, that she found them in the park.'

'Right…' Nath says as if this isn't important.

'If that's true,' I say, 'it means your mum lied.'
'Lied?'
'Your mum said the police found Abigail abandoned in the park, that Natalie had left her on her own.'

Nathan nods, as if remembering.

'But what if that's not true? If your mum found Natalie before the police, it means she had a chance to stop her – if she saw her that day with Abigail, then she let her go and made Abigail lie to the police about it.'

'No,' Nath says, shaking his head decisively. 'She wouldn't do that. Why would she do that?'

'I don't know, Nath. But there's obviously a reason she didn't want Natalie caught—'

'Are you saying my mum is involved in all of this?' he asks, his cheeks colouring as he stands up, moving away from me. 'Because she's not.'

'Calm down,' I say, shocked at the anger in his voice. 'I'm just saying we should talk to her before we call the police – find out why she let Natalie go in the first place.'

'You can do what you like,' he says, 'but I'm going to look for them both. Now. You can't bring a car up to the venue, not without authorisation...' He pauses, and I nod, remembering the fuss this caused with getting all our guests' number plates ahead of time so they'd be allowed to drive up to the reception. 'So if Natalie has come for Abigail, she must be on foot.'

With that, Nathan takes off his jacket and dumps it on the chair beside me, then walks out of the front doors. I press my fingers into my temples; this is not how I thought things would go. How could I be so stupid, leaving Abigail like that? Why did I trust the word of a five-year-old little girl? Every single decision I've made since I went to look for Natalie has been the wrong one. I've brought this all on myself.

'Sally?' Amanda's voice breaks me out of my trance, and I open my eyes. She's sitting in front of me, crouched on the floor. 'What's wrong?'

Her kind eyes, the eyes I've looked into since I was a child, are the thing that breaks me and the tears come, thick and fast.

Despite my earlier suspicions about Amanda's lies, I fill her in without hesitation – as much as I can through the choked sobs. To Amanda's credit, she reacts exactly as I'd expect her to – calm, collected, commanding. She tells me not to worry, she'll call it straight in and they'll find Abigail – again.

'But Amanda...' I say, having regained my composure a little. 'Abigail said you spoke with Natalie, before. The last time she took her.'

Amanda frowns, her head to one side. 'What do you mean?'

'I don't know,' I say, helplessly. 'It was just something Abigail said earlier. That you told Natalie to stay away, but that doesn't make sense because Natalie abandoned her in the park in Bristol, didn't she? You were here, in Stratford, when the police found Abigail, weren't you?'

Amanda nods. 'Yes, Abigail was left alone in the park, and I was at home, like I said. If I'd been within ten miles of that woman, I'd have made sure she got brought in and arrested. Trust me. She needs to be behind bars, that one.' She says it so confidently that I wonder if I even heard Abigail correctly earlier. Maybe she didn't say that at all. She is five, after all, and her stories often make no sense whatsoever.

'Where's Nathan?' Amanda asks.

'Gone out to look for Natalie,' I say. 'I didn't want to call the police...'

'Whyever not?' she barks.

I shake my head, feeling stupid for getting so carried away with Abigail's story. 'What Abigail said about you warning Natalie off in the park, it just made me think... that you lied and if you did, you must have had a reason to...'

'*Lied?*' Amanda says, her hand coming up to her chest. 'What would I have to lie... Look, never mind all of this. I'm calling it in now.' She turns her back on me and fishes her phone out of her beige clutch; I wonder what she imagined she'd be using this bag for when she dressed this morning. Topping up her lipstick, fetching the tissues during our speeches, anything except reaching for her phone to dial in another child abduction. She steps outside and I watch as she makes the call, then she disappears from sight.

'Sally?'

I turn around. It's Chloe, her chest sprawled with a red rash.

'Sally, have you seen Abigail?' Her words are calm, but her expression is anything but. 'I can't find her. She was playing out here and now I can't find her.'

'It's okay,' I tell her, knowing there's no point worrying her at this stage. 'Amanda has taken her outside to play.'

'Oh, good,' Chloe says, flopping herself down on the chair next to mine. 'I was so worried. God. That feeling, you know? I don't think I'll ever get over that feeling.'

I nod, guilt creeping through me.

'Could you go grab me a glass of prosecco?' I ask Chloe, desperate to be away from this conversation. 'I'll meet you at the bar in a second. I just need to pop to the bathroom.'

'Oh,' Chloe says. 'Yes, no problem. See you in a sec.' She looks at the front door, as if she'd much rather go outside and find her daughter with Amanda, but she's too polite to refuse my request and after a second, gets up and heads to the bar.

The moment she's gone, I rush outside.

'Amanda?' I call but there's no reply. 'Amanda?'

I scan the area around me, expecting to hear sirens blaring as the police make their way to us, but it's quiet except for the sounds of the melodic birdsong high in the sky above us. The sun is out – the perfect day for a wedding.

But I don't hear any of that, because cutting through all of it, there is a scream.

CHAPTER FORTY

I run without thinking, without waiting for the police, without being sensible. My wedding dress is heavy and impossible to move in and yet the wind whips through my hair, setting it free from the pretty plait that was all I cared about a few hours earlier.

I follow the sound of the screams, which have become more frequent – words I can't make out through the sound of my blood pounding in my ears – until I reach the entrance of the woods.

'Nathan?' I shout hopefully. 'Amanda?'

Neither answers and I step forward into the trees, dragging sticks and leaves with me, trapped in the tulle of my wedding skirt. I listen again but everything has fallen silent. The police will be here soon; Amanda has called them. Everything is going to be okay. I keep moving forward, ignoring the sounds of my beautiful, once perfect, dress ripping as it drags on the forest floor. Our wedding planner sold these woods to us as the perfect backdrop for wedding photos, never intending for us to go into them. The sunlight is choked in here; the tall trees sway in the light breeze high above me but I can no longer hear the birds singing.

'Get *off* my Manda!' Abigail's high-pitched voice shrieks out across the forest, sending a shockwave through me.

I take off in the direction of the sound, heading deeper into the trees, and a few minutes later, there they are in a small clearing. I stop dead still behind a tree, afraid that moving might break the picture in front of me – might make things even worse.

Abigail is wrapped around Amanda's legs, looking over her shoulder and scowling at Natalie, who stands a few feet away.

'Gracie, come on. Come to me. I'm your mummy,' Natalie pleads, her voice hoarse as if she's been screaming, desperation in her eyes as she gestures for Abigail to come towards her.

'You're not my mummy,' Abigail says. 'And my name's not Gracie!' She turns her face away and buries it in Amanda's stomach.

'You should go now,' Amanda says, her voice like ice, but Natalie only shakes her head.

'I'm not doing that again,' she says. 'I'm sick of running. Where has it got me? I've lost everything! I want her back.' Her voice breaks and despite everything, my heart breaks for her as she utters a final plea, 'I want Gracie back.'

Amanda shakes her head, unmoved. 'She's not yours to have, Natalie.'

'She's not *yours*!' Natalie shrieks, stepping towards Amanda.

Abigail screams again and unwraps herself from Amanda legs, then pushes Natalie in the shins. It's not enough to hurt her but Natalie looks wounded as she takes a step back, tears in her eyes at the betrayal – the rejection.

'See, Natalie? Abigail doesn't want you,' Amanda says, her words so cold it sends a chill through me as she takes

Abigail by the hand. 'You're not good enough to be her mother. You never were.'

'You never gave me a chance to be,' Natalie spits back. 'Why did you do it? I don't understand why you had to do it.'

'Oh Natalie,' Amanda says. 'Of course you do.' She turns to walk away, picking Abigail up on her hip. 'I'm going to offer you the same choice I did last time, and this time I think you should take it. Continue to stay away and nothing bad will happen to you. The money will keep coming. The agreement will stay in place. But this is the last time I give you this opportunity, Natalie. Your last chance. Next time, I won't be so generous.'

I'm trying to work out Amanda's words in my mind, the sentences making sense yet adding up to something incomprehensible to me – when Natalie pulls a knife from the inside of her coat.

'Put down my child or I'll bury this so deep in your chest they won't even be able to pull it out,' she says, her voice no longer wobbling but firm, stable, *terrifying*.

'Stop!' I shout, coming out from behind the tree.

Both of them startle, staring at me with wide eyes.

'Amanda, put Abigail down,' I say and Amanda's eyes flit to Natalie.

'Natalie, can you put the knife away so Amanda can put Abigail down?'

Natalie nods, taking a step back from Amanda and lowering the knife.

'Abigail, sweetie, come to me,' I say as calmly as I can and, for once, she does as I ask. The second she's near enough, I grab her in for a hug; her body is hot to the touch.

'I know you care about your daughter, Natalie,' I say. She scowls at me. 'And so I'm going to ask you to let me send her back to the hotel where she'll be safe.'

Natalie shakes her head, her face pained. 'No. She has to stay here, with me.'

Amanda opens her mouth, but I shoot her a look that tells her to keep quiet. Against my legs, Abigail starts to cry.

'Natalie, look,' I say, nodding towards the little girl. 'I'll make sure you see her again, but please. Let her go. She doesn't need to be here.'

Natalie looks from me to Abigail, her pained sobs bouncing between us as my heart bashes against my ribs. Finally, Natalie nods. I don't waste a second, dropping to my knees to come face to face with Abigail.

'I need you to be a big girl and go back to the hotel, okay, sweetie? You go that way, and don't stop until you get there, okay?'

Abigail nods up at me, her eyes wet with tears before I give her a little push in the right direction.

I watch her run, her pretty dress puffing out like a princess, until she disappears from view.

'Natalie, put the knife down,' I say but she shakes her head. 'We can talk about this. You don't need to do anything violent.'

She laughs, a sharp spiteful sound as Amanda takes a step towards her.

'I said don't fucking move!' Natalie is quick on her feet – lurching towards Amanda and grabbing at her arm as Amanda trips, the heels that she complained about earlier failing her as she twists her ankle. Pain ripples through her features but she contains it.

'I'm not leaving again,' Natalie says. She twists Amanda's arm behind her back then she presses the knife against her throat. 'Not without Grace. You can't make me.'

She's small next to Amanda but there's a fire in her eyes that tells me she won't let this go without a fight.

'I don't want to make you, Natalie. We all just want to sort this out—' I say but Natalie cuts me off.

'No, you don't!' she screeches. 'You're a fucking liar! You're just like *them*.'

'Like who?' I ask as the sound of a stick snapping underfoot from behind me makes me twist around. 'Nathan,' I say, rushing into his arms. His face is pale, his eyes round as he takes in the sight of his mum held by Natalie, a knife at her throat.

'Mum,' he says, his voice breaking on the softness of the word.

'Oh, here he is,' Natalie sneers, her eyes burning. 'Here to save the fucking day as usual.'

I snatch a look at my husband. His face is contorted, and I know then he has not been honest with me. That my dad was right. I take a step away from him.

'Natalie, please,' he says. 'Let her go. We can talk about this.'

'Talk about what, *Nathan*?' She spits his name like it is a lie and he bristles. 'How you promised to save me and instead ruined my entire fucking life?'

'I'm sorry, Natalie,' he says, taking a step towards her, his voice back to the steady tone I've heard him use so many times on work calls. 'I know what happened wasn't right—'

'Wasn't right? You took my baby!'

'No,' he says. 'You *left* your baby—'

'She told me to!' she shrieks, pointing at me, briefly letting go of Amanda's arm but then grabbing it again. 'You told me to leave her, you promised me she would be safe. But you were setting me up, you were all setting me up.'

'No,' I say, shaking my head. 'I promise you, Natalie, all I ever wanted that night was for Abigail – Grace – to be safe.' I lock eyes with her, begging her silently to believe me but her gaze is lost as she looks back to Nathan and I see it then – her broken heart.

'What did you do?' I say to him, twisting from his arms. 'What did you *do*?'

CHAPTER FORTY-ONE

'Sal,' he says, 'please.' He holds out his hand, his shiny new wedding ring bold against his skin but I shake my head.

'Are you going to tell her, Nathan,' Natalie spits, 'or shall I?'

Amanda looks at Nathan and I watch something pass between them, the pit of my stomach swirling.

'She already knows I was undercover,' Nathan says and Natalie smirks.

'So, you told her about us?' she says and I hear myself taking in a sharp breath, my worst fears coming true.

'There was no *us*, Natalie,' Nathan says, and Amanda has the good grace to at least look ashamed at the bare-faced lie from her son.

'You tell yourself whatever you need to, *Nathan*,' Natalie says, his name sounding like a lie on her tongue. 'But you know what you promised me. You know the lies you fed me – you said you'd take me away, keep me safe. You swore that to me.'

I look at Nath, his eyes now glued to the forest floor as Natalie's heart lays broken between us.

'But you didn't, did you?' Natalie spits, her words taking a different shape now.

Nathan flicks a glance over to his mum, Natalie's knife still pressed into her throat. He opens his mouth to speak but Amanda starts to shake her head. Natalie tightens her grip and Amanda winces. Things become very still.

'The two of you,' Natalie says, her voice painfully low. 'You murdered Shane and stole my baby.'

'No!' Nathan says, stepping forward towards her but I grab him when I see the fury in Natalie's eyes, sharp as the edge of the knife in her hand.

'What does she mean?' I ask him. 'What did you do?'

'They set me up!' Natalie shrieks, momentarily removing the knife from Amanda's throat as she points wildly at me instead. 'Like I told you!'

'It was an accident,' Nathan says, his voice so small and childlike that I do a double take.

'What was an accident?' I say, turning my back on Natalie so I can look him in the eyes.

Behind us, there's a commotion and I whip around.

Amanda has broken free from Natalie's hold and now twists Natalie's arm behind her back. She screams out in pain but it's no use – Amanda is in full control. Natalie's face contorts in rage as she struggles, uselessly, to break free as the knife lays on the forest floor between them.

'Nathan,' Amanda commands. 'Help me.' Amanda drops Natalie to the ground, her arm pressed against Natalie's back as she pushes her face down onto the ground.

Nathan steps forward like a robot and places his knee on Natalie's back, taking over from his mum, holding Natalie in place. She flails around on the floor, fighting, but I know she won't be able to wriggle out of this. I've seen this move on officers' bodycams hundreds, thousands, of times. It can render the most violent, enormous criminal into a helpless toddler. I take a second to calm myself down and ignore the urge to help Natalie, as she lies defenceless amongst the twigs on the dried muddy floor. This is better – this is contained. Things are going to be okay now.

'Where are they?' I ask, watching Natalie on the floor. Amanda looks at me, her eyes cold. 'Amanda, where are the others? You called it in. Didn't you? Back-up will be here soon.'

Amanda sighs and Nathan looks at me, his big eyes trying to communicate something I can't – don't want to – understand.

'No one is coming, Sally,' Amanda says. 'We have to deal with this on our own.'

Natalie turns her face towards me, sticks digging into her pale cheek, tears falling helplessly from her angry eyes.

'What do you… What do you mean?' I ask, my voice shaking.

'Sal,' Nath says, his eyes pleading. 'You should go.'

'Go?'

'Please—'

'What is happening? Get off her.' I pull at his shoulders but he's unmovable and Amanda guides me away. 'What are you going to do to her?' I ask, whipping around. 'What are you going to do to her?'

'They're going to kill me,' Natalie manages to say. 'Just like they killed Shane.'

The woods around me swell, the darkness becoming all-encompassing as I take in Natalie's words properly for the first time. *They killed Shane.* No. This can't be right. I look at Nathan, his knee pressed firmly into Natalie's back, but he won't meet my eye. I turn to Amanda who looks straight through me. Panic tears through my body.

Nobody is coming.

Amanda didn't call for back-up because she never planned on letting Natalie leave these woods alive.

'We need to be quick,' Amanda says to Nathan, unruffled by Natalie's bombshell. 'Abigail will tell someone at the party. We don't have much time.'

'Much time to *what*?' I ask.

'You should go, Sally. Go back to the party.'

'Mum,' Nathan pleads.

'There's no other way, Nath,' she says calmly and a cold chill runs through my entire body. 'You know what we have to do.'

Nathan closes his eyes and takes a deep breath as Natalie squirms beneath him.

'This is insane,' I say. 'You can't be serious. What are you going to do? Amanda, what are you going to do?'

'Go,' she says again but I shake my head, planting my feet firmly on the floor – an immovable object.

'I need to know what you did,' I say. 'I'll do whatever you want me to. I won't say a word. But first, I need to know.'

Amanda looks at me, considering this. We hold each other's gaze, something passing between us as I look at the woman I have spent my entire life admiring, and wait for the image I've held of her in my head to be set on fire.

'Get her up,' Amanda tells Nathan, not taking her eyes from mine.

He stands Natalie up and I wince at the scratches on her face, a stray twig cutting into her cheek. He holds her arms firmly behind her back and pain contorts her face. She looks childish, helpless. This is no master criminal; no vindictive, heartless mother.

'She can tell me,' I say, pointing at Natalie.

'She's a liar,' Amanda says. 'You know what sort of person she is, Sally. You can't believe a word she says.'

'I don't care. She can tell me, or I go back to the wedding right now and get every officer in the room down here.'

Amanda looks at Natalie and nods.

'I didn't kill Shane,' Natalie says. 'I told you that on the day we met.'

'Sally,' Nathan says, pain in his eyes as he looks over Natalie's head at me. 'I'm sorry, I'm so sorry...'

Realisation dawns on me then and I stumble backwards. 'No. You didn't. Nath, you didn't...'

He lets go of Natalie and comes towards me, grabbing me by the wrists. 'I'm sorry. I didn't mean to... It was an accident, I—'

'You didn't kill Shane,' Natalie's voice rips through us and we both turn to face her. 'Is that what you think?' she says, a confused frown on her smudged face.

'I did,' Nathan says. 'I hit him, and he fell and...'

Natalie laughs, a cruel cackle that rings out through the woods. 'Fuck me,' she says once the laughter has stopped. She turns to Amanda. 'You really are a cold-hearted bitch. You never even told *him*, your own son.'

Amanda sighs, like this whole conversation is beneath her.

'Mum?' Nathan says, his voice so much like that of the little boy I grew up with.

'You're so fucking stupid,' Natalie says to Nathan. 'You didn't kill Shane. She did.' Natalie points to Amanda and Nath takes a heavy step back, bumping into me, knocking me over so that I don't see what happens next but when I stand up, Amanda is on the ground and Natalie is on top of her, the knife back in her hand wavering near Amanda's throat as she talks at Nathan.

'She suffocated Shane,' she says, nodding at Amanda. 'After you left. You knocked him out, yeah, but he would have been alright. But then this bitch came along, promised to help me, then held her hand over his fucking mouth.' Natalie looks down at Amanda then acts the horrible scene out, her hand blocking the air from Amanda's mouth 'Didn't give a fuck as she watched the life drain out of him, did you?'

'Stop,' I say, stepping forward. 'Stop, you're killing her.'

Natalie lifts her hand from Amanda's mouth as she splutters, her eyes cold with hatred staring at Natalie.

'He was *scum*,' Amanda spits. 'Shane Blackwell was scum.'

'But Mum… How could you?' Nath says. 'All this time, I thought I'd killed him. You let me think I'd killed him.'

'What other choice did I have?' Amanda says.

Nathan lurches forward in a rage and for a second I think he's going to hurt Amanda but instead he rips Natalie from her position seated on top of Amanda's chest and pulls Amanda up to standing. Natalie appears so shocked at the quick intervention that she doesn't fight it and there's a moment of silence in the woods as we all wait for what will happen next as the knife glimmers in the sunlight, equal distance from us all.

Natalie comes to her senses and suddenly tries to make a run for it, but I grab her.

'You're staying,' I say, gripping her arm hard beneath my fingers. 'I'm not letting you run away again.'

Natalie looks up at me with hurt in her eyes, but I ignore her silent plea. I need her to stay for this to make sure Amanda tells the truth.

'What happened that night?' I ask Amanda. 'Tell us everything, now. Or I'll take Natalie back up to the party myself and hand all of you in.'

CHAPTER FORTY-TWO

THAT NIGHT, FIVE YEARS AGO

AMANDA

Amanda is already out when Nathan calls; on her way to Tesco for a post-shift bottle of wine.

'Mum?' he says and instantly she knows by the sound of her son's voice that he is in trouble. 'I need you to come.' As he reels off the all-too-familiar address she has an overwhelming sense of foreboding, instantly, that this is going to be the end of their life as she knows it.

She had known from the moment she discovered that his girlfriend – that nasty little drug addict Natalie Pierce – was pregnant. Nathan hadn't told her he was working undercover, but she knew. She let him think he was doing it without her approval – without her involvement – but of course the ACC of West Mids, a long-time friend of hers, wasn't going to put her son into an op without giving her the heads-up.

So she'd watched on carefully as her son became entwined with drug dealers and sex traffickers, all the time looking and playing the part of a young, innocent kid with no idea what he was doing. Which was ironic, really, given that he really did end up so out of his depth with the situation.

She'd noticed Natalie early on, too. She was young – too young to be caught up in a life like this – but Amanda had been doing her job for long enough to know girls like Natalie always end up in these situations. She was pretty, slight, dainty. The sort of girl that if she'd had money might be the queen bee of her private girls' boarding school, her biggest concern whether Claudia or Cassidy might overtake her in the popularity contest one day. Amanda wasn't worried about Nathan and Natalie at first; she knew her boy – she'd brought him up right. But then she heard whispers of them getting a little too close, a little too friendly. When Amanda saw Natalie on the cameras one day, the swell of her stomach a sickening, disgusting sight, she knew she had to act.

Amanda pulled the undercover operation the second she found out, taking no chances, citing flaws in the operation as a reason to get West Mids off her patch. It was easy to do; as ACC she had control over what went on in Warwickshire, even if it wasn't her operation. She never told Nathan what she'd done. He believed she still had no idea he'd been working undercover on her doorstep for months and didn't know the op getting pulled was anything to do with her at all. She'd done it to protect him – to keep him as far away from Natalie Pierce as she could until she figured out what to do about that baby – her grandchild.

Now, Amanda arrives on the scene where Nathan has called her. She made the journey in ten minutes, putting her foot flat to the floor. She's already told her son to disappear – to get himself an alibi, immediately. She'll tidy his mess up, she always has.

The dark, abandoned car park is quiet – too quiet. She knows the sound of sirens will flood the scene within minutes, so she doesn't have long.

'Natalie?' she calls out as she steps over the hard gravel. 'It's okay. You're safe. You can come out.' She holds up her badge, high above her chest and moves it in a three-sixty-degree motion.

It works.

The girl creeps out from behind the bins, her small frame shaking. Amanda watches as her body crumples in relief at seeing the badge Amanda holds up like a beacon.

'Stay there,' Amanda commands her. 'Chuck your phone across. Now, please.' Natalie looks doubtful but then does as she says, throwing her phone across the carpark and landing it at Amanda's feet. 'Where is he? The injured man. Where is he?'

Natalie points to the left of Amanda and she sees him then, or at least, his feet poking out behind the wreckage of what used to be a scrapyard.

'Stay there,' she commands Natalie, then walks over to Shane.

The blood is pooled out around his head. Nathan told her it was one punch, that Shane fell down *like a sack of shit* and smacked his head on a cracked paving stone. Amanda inspects the wound and then hears it, a low, desperate groan.

The man's eyelids flutter.

Amanda sighs.

She looks at her watch. She has maybe two minutes; maybe less.

She sits on the man's chest – his breath escaping from his mouth as his eyes go wide. He looks at her, those shark-like eyes boring into her soul. She knows this man;

he is not the kind of man she is going to shed tears over. Shane Blackwell is a cockroach. They have been trying to stop his drug and sex trafficking operations for years, but they've never been successful. He rarely gets his hands dirty, preferring to use girls like Natalie, barely out of school, to do his business for him. The world will not miss him. No one will miss a man like this.

Amanda tells herself this as she presses her gloved hand down over his mouth, then takes her other hand and holds his nose. She doesn't flinch as she watches the life spill out of him, his shark eyes turning darker as he leaves this earth. If she believed men such as this could have souls, she would say she sees his float straight out of him – but men like him don't have a soul. Men like him are nothing. She stands up, presses her foot into his ribs to check there's nothing left, then turns away, satisfied the job is done.

'What did you...'

The girl – Natalie – has not followed her instructions and is shaking in front of her.

'What did you do?' she says again.

Amanda sighs, annoyed that this stupid girl can't learn when to keep her nose out.

'You killed him,' she snivels as if the man in question hasn't made her own life a living hell for the last five years. As if she isn't glad to see him dead.

'Natalie,' Amanda says calmly. 'I need you to listen to me. In less than one minute, this entire scene is going to be swarming with police. If you want to get out of here, you need to do it now.'

'But you *are* the police... What's going on? Where's Lee?'

'Don't worry about Lee,' Amanda says, not flinching at the use of her son's undercover name. 'I'm giving you the opportunity to get out of here – if they come, you'll be arrested. Do you understand?'

'But I haven't done anything,' Natalie says, her eyes glued to Shane's dead body on the floor.

Amanda waits a beat, ignoring the countdown clock in her head. 'I know that. But that's not how it looks, is it? I want to help you. I promise. You need to go, now.'

Natalie nods, then moves back towards the bins where Amanda knows the baby is.

'No,' she says. 'Leave her.'

'Leave her?' Natalie says, whipping around.

Amanda nods. 'She'll be safe. I'll bring her back to you, but they'll catch you if you take her. And you'll lose her forever.'

Natalie opens her mouth as if to speak and Amanda sees her brain trying to work it out.

'You can trust me,' Amanda says. 'I won't let anything bad happen to you – to either of you. Go to your flat – I'll be there in an hour with your baby—'

'Grace,' Natalie says. 'Her name is Grace.'

Amanda nods; it's a nice name. The sort of name she might have chosen for her own daughter if she'd ever had one. It will have to go, of course.

'I'll bring Grace to you,' she says. 'Safe, I promise. But you have to go.'

Natalie looks conflicted but the sound of a siren approaching spurs her into action.

'You promise you'll bring Grace back to me?' she says, tears pooling in her mascara-smudged eyes.

'I promise,' Amanda says.

With that, Natalie takes off.

Amanda walks to the bins, expecting the sound of cries.

But her granddaughter is silent, her beautiful big brown eyes searching her surroundings as Amanda picks her up in a bundle.

'I'll keep you safe, darling girl,' she whispers to the tiny, innocent baby in her arms. 'Grandma's here now and I promise, I'll always keep you safe.'

–

'What's happened here?' Amanda asks, fifteen minutes later, pulling back up to the scene as if for the first time, and getting out of the car. A young detective she doesn't recognise pulls himself up to his full five-foot-eight and sticks his stubbled chin out. 'This is a crime scene. You can't be here.'

'Pipe down,' she says, walking past him and flipping her badge.

'Sorry, Ma'am,' he says, his voice wobbling at his mistake. 'I didn't…'

She waves her hand at him and steps over the crime scene tape.

'Ma'am,' the other officers chorus as she approaches.

'Are you here as Gold Command? I thought Chief Super Jones was on his way,' the most senior officer asks her, rather boldly – but she lets it go.

'I was on my way home,' she says, eyes scouring the scene. 'Thought I'd see what all the fuss is about.'

'Shane Blackwell,' the officer says, pointing to the dead man's body to the side of them. 'Got called in about half an hour ago – looks like a fight gone wrong. But Ma'am,' the officer says, his voice catching. 'There's something else.'

'What is it?'

The officer beckons her around the corner, to the bins.

'The perp, she left her baby.'

There's a female officer holding Amanda's granddaughter tight to her chest, the disgust on her face visceral that someone could have left something so small and innocent in this wreckage.

'Jesus,' Amanda says, stepping forward and reaching for the blanket that covers her precious granddaughter. 'What sort of mother leaves her baby like that?'

The other officer shakes her head. 'The call handler told her to keep the baby hidden, apparently. But she didn't expect the bloody woman to run off and leave her altogether.' She sighs, 'Some people don't deserve to be parents.'

The baby looks up at Amanda, her eyes growing wide. 'Looks like she likes you, Ma'am,' the officer says and Amanda smiles as she passes the child to her.

Amanda takes the warm bundle in her arms, smelling the sweet innocent scent of her granddaughter. Since she'd found out Natalie was pregnant, she has waited for this moment. There was no way she could let this baby grow up in a life like Natalie's. What Nathan had done, starting a relationship with a young, vulnerable woman while undercover, was despicable, but how could Amanda blame him when she knew what girls like Natalie were like? She'd worked with them her whole career. Ruined, early on. She shouldn't say it – shouldn't even think it – but it was the sad truth of the world. There was no helping girls like Natalie.

Amanda knew from the moment she saw Natalie's swollen pregnant belly that she'd never let her grandchild grow up to be a girl like that. No matter what it took.

No matter what.

Later that night, her grandchild safe in the care of social services until they knew what to do with her, he calls her.

'Amanda?' His voice is tired, annoyed. 'It's done.'

She releases some of the tension she's been holding in her shoulders since she took that call from Nathan a few hours before – though it feels like days.

'Thank you,' she says down the line.

'I don't like this,' he says.

She sighs. 'I know. But what else could I do, Graham?'

Chief Superintendent Graham Jones had been the only person Amanda could count on to finish the job for her tonight. He was her oldest friend, the only consistent man in her adult life. She hadn't told him the whole truth – he would never have helped her if he thought she was involved with Shane's death – instead leading him to believe Nathan had accidentally killed the man and Amanda had covered it up.

He also knew about the baby; he knew she was Nathan's. He understood why Amanda wanted to get rid of Natalie rather than let the courts take their own decisions about the care of this baby that was so precious to Amanda, but he didn't agree with it. Still, he'd done his part – turning up to meet Natalie in Amanda's place and paying her off to disappear, for good, warning that if she ever tried to come back, they'd make sure she was locked up for life.

'Sally would be ashamed of me,' she hears Graham mutter down the line now.

'Sally will never know, Graham,' Amanda tells him, impatient now that he is clearly struggling with their decision. She needs him to be the strong, stable man he usually is. 'I promise you, no one will ever know about this.'

She knows it's wrong to let the two men in her life – her son and her oldest friend – believe that Nathan killed Shane. But if Nathan knew the man was still alive when she turned up, he'd have made her do the right thing and let him live. And where would that have left her son? He'd be facing charges for aggravated assault on top of corruption – after all, Nathan was going to take Natalie away tonight and she *is* a criminal. He'd lose his job, and quite possibly end up in prison. And she knew what happened to police officers in prison. She'd never let that happen to her boy.

The baby is safe now. She'll be looked after, adopted by a good family. Amanda would stay close – forever – watching over her grandchild.

Natalie would stay gone – paid off every month to disappear – and everyone could be happy.

Nathan's involvement would never come out.

Amanda can live with the guilt. It's easy to do the wrong thing when it's for the right reasons. And if Graham couldn't – if for any reason his moral compass faltered – she'd remind him what he did for his own daughter all those years ago.

CHAPTER FORTY-THREE

SALLY

'My dad did... what?'

Amanda's admission is like a knife to the gut. Though I've suspected it from the moment I saw the money going out of my dad's account the night of Shane's murder, hearing it confirmed is something else. How could he do this to Natalie? How could he do this to me?

Nathan is silent, staring at his mum with a heartbroken expression on his face and something else I can't name.

'He was helping me,' Amanda says, sticking her chin up like she hasn't just told us the most damning story you could imagine. Like she's proud of what she's done. 'He knew the best thing for Abigail was to get her out of that house and away from—'

'The best thing for...' Nathan starts with a bang but can't find the words to complete his sentence as the tension hangs in the air. 'Mum, you...'

He shakes his head and reaches towards me, but I shake him off.

'Don't touch me,' I whisper, looking down at my torn and filthy wedding dress. I'm married to a man who got a young girl pregnant when he was supposed to protect her; I'm married to a man who believed he killed someone and let the mother of his child take the blame.

'Sally,' he says, pleading. 'You don't understand—'

'I understand just fine,' I spit. 'The two of you did something despicable that night – but I think I could have forgiven it. You made a mistake, Nath, you fell in love with the wrong person' – he shakes his head – 'let me finish. And you lost your shit with Shane. I even understand you calling *her*' – I point fiercely at Amanda – 'to help you when shit hit the fan.' I take a breath, tears threatening. 'But what I don't understand, what I will *never* understand is how you forced Natalie into hiding for *five years*. You made a bad decision – but you've chosen to make it again and again and again every single day since.'

Amanda shakes her head. 'Your dad understood. He knew what women like that did to children if they stayed in their lives.'

'What do you mean?' I snap.

She looks at me, pity in her eyes. 'Come on, Sally. You know exactly what I mean. Your dad did what he had to do with your mother, and so he understood why I had to keep Natalie away from that child.'

'My mother... What do you mean? My mother left.' I drop my hold on Natalie and take a step back into the woods, the twigs cracking beneath my feet. Natalie takes a step towards Nathan, but he holds his hand up to her, stopping her from leaving.

Amanda looks at me, saying nothing, and the implication of her silence hits me so hard I have to take another step backwards.

'My mum didn't leave me,' I say, almost to myself. 'My dad forced her to go.'

'Sally,' Amanda says, her voice stern. 'I know this must be hard to hear – and I understand you don't want to believe that your dad did what was best for you and then later, what was best for Nathan and Abigail, but you must believe me when I say that everything I did, I did for Nathan and that tiny baby. When it comes down to it, we will all do horrible things if it means protecting our own flesh and blood.'

Nathan laughs, a sick, rasping sound that makes us both stop and turn to him.

'Fucking hell,' he says, his face turning white. 'This is so fucked. This is so… Mum, Abigail isn't your "flesh and blood". She is *not* your grandchild.'

'What?'

Nathan shakes his head. 'I didn't have sex with Natalie,' he says, with such distaste I almost believe him. 'Yes, I got too close. I made her promises I couldn't keep – but I never touched her. *Never.*'

Amanda shakes her head, unable or unwilling to believe this and I look at Natalie who has remained silent throughout our latest exchange.

'Is that true?' I ask her and her eyes meet mine. For a moment, I hold her gaze and beg her to remember the friendship – odd and brief though it's been – we shared.

She looks between us all, like she's weighing up her options. Then nods.

'You and Nathan, you didn't sleep together?' I ask again.

'No.'

'The baby…'

She shrugs. 'Probably Shane's. Or one of his mates'. I don't know. But she's not Nathan's.'

I look at Nathan, his shoulders slumped like a man getting ready to serve a life sentence and I'm filled with a rush of love for him. Of course he didn't sleep with Natalie.

Out of nowhere, Natalie lets out a laugh. She's looking straight at Amanda's white, horrified face.

'You dumb bitch,' Natalie says through the laughter. 'All this time, all this effort, for a baby that's nothing to do with you.' Her words are cruel, but they are fuelled by years of hatred and I don't blame her for them.

'No,' Amanda says. 'You're lying. Both of you, you're lying. Abigail is my granddaughter.'

Amanda's fists clench either side of her body and I imagine all the memories spooling through her mind right now. All the hours, days, weeks, years she's spent with the little girl she loves so much; the little girl she believed was her blood. The pain on Amanda's face is almost too much to bear but it doesn't deter Natalie.

'Grace is nothing to do with you,' she spits. 'And I'm going to make sure you never step within a hundred feet of her when this is done, you mad bitch.'

Amanda springs into action so fast that none of us see it coming and suddenly she's on top of Natalie, screeching as she claws at her face like a feral cat. Natalie is small but wily and quickly flips her hips up to bounce Amanda off her in a move so violent that I wonder if Amanda will ever get back up from the forest floor, but she does – springing up in a blur, until Natalie runs at her full force, tackling her like a rugby player and knocking the wind from Amanda's lungs.

'Stop!' I shout, grabbing at Natalie as Nath grabs his mum but neither of us are successful at holding them back from each other and suddenly Natalie is back at Amanda's throat, her hands choking the air straight from her.

Amanda gasps for breath, her eyes pleading with Natalie to stop as I watch the life slowly draining from the woman who has been there for me my whole life.

I move without thinking, grabbing the closest thing to a weapon I can find, and hit.
Hard.

CHAPTER FORTY-FOUR

Everything falls silent.

I look at the jagged rock in my hand. It's sticky. Wet. Covered in...

'Oh my god,' I say, dropping it from my hand. 'Oh my god.'

Amanda stands up then picks up the bloodied weapon. 'It's okay,' she tells me.

'It's not okay, she's...'

I look at Natalie's lifeless body on the woodland floor in front of me, sticky red blood pooling around her head. Her face is white and her eyes flicker, then stop.

'Is she dead?' I say. The words sound like something from a devastatingly bad script and a spurt of laughter bubbles in my throat. It erupts – quickly and without warning – then is smothered by my uncontrollable sobs.

On the floor, Amanda is by Natalie's side, taking her pulse, listening for a breath. She shakes her head at Nathan who nods, stoically.

'She's okay,' I say to them. 'Isn't she? She's okay – she's just hurt. She's okay, isn't she? Will you please tell me she's okay. Please,' I beg, but neither of them respond, looking at each other like they're forming a silent plan.

'Sally,' Amanda says, standing up in front of Natalie's body, blocking some of my view. 'Everything is going to be okay. But I need you to calm down.'

In the distance, we hear sirens.

'Oh my god,' I say. 'Abigail must have told the others; they've called the police. They're going to find her – they're going to find us—' I speak without thinking, without connecting that what I'm really saying is they're going to find me. A killer. *Me.*

'It's okay, sweetheart,' Amanda says, holding out her hand to me but not stepping forward as I spot the red tinge of blood around her nails. Natalie's blood. 'Everything is going to be okay.'

Nathan grips me from the side, his arms around my waist the only thing stopping me from falling to the floor where Natalie's lifeless body lies.

Amanda coughs, rubbing at the red marks around her neck – the ghost of Natalie's angry fingerprints embedded into her wrinkled skin.

'I hit Natalie,' Amanda says, staring at me with those famous ice-cold eyes.

'What? No. I hit her. I was trying to get her off you—'

'No, sweetheart,' Amanda says again, firmer this time as she grips my arm. 'You were stood right there, right where you are, when you saw me pick up this rock from the floor beside me and hit her with it.'

'No,' I say, openly crying now as I begin to understand what Amanda is doing for me. 'No, you can't.'

'Mum,' Nathan says, his voice like a little boy's. 'We can fix this – you don't need to take the blame.'

'No darling, there is no fixing this. What's done is done. And it's my fault, all of this, it's my fault.'

'You were just trying to protect me,' Nathan says but Amanda shakes her head, dropping my arm to move closer to Nath.

'But I was wrong, wasn't I?' she says. 'I should have known you'd never get that girl pregnant.' Her voice catches on the end of her words, and I think of Abigail – the grandchild Amanda has lost who was never hers to begin with.

I shake my head as my vision blurs, but Nathan holds me tight, not allowing me to wriggle free of his grip. I get the mad urge to run, to abandon them both and flee through the forest in my torn wedding dress like something beautiful and awful.

'We need to get our story straight,' Amanda says as the sirens draw closer and my feet plant more firmly on the floor. There is no fleeing this. No escape.

'We can't do this. I can't let you take the blame,' I say but no one is hearing me. Not any more.

Nathan and Amanda talk, his eyes wide and attentive as he listens to her instructions the same way he did when we were kids, and she was explaining the rules of a new board game. Nathan has always been good with instructions.

Soon, the instructions are over and it's time to act.

My instructions are clear. Go back to the party. Raise the alarm. Wait until the police arrive and tell them where Natalie is. I am in shock, Amanda tells me, remember I am in shock and can't immediately explain what has happened. She will give the police the story we have agreed, and later, when my shock wears off, I will repeat the same story. As will Nathan. Amanda repeats my instructions again, like an army commander and I feel myself nodding, agreeing to this awful plan as if I have no other choice.

'Go,' Nathan says, softly, followed by a firmer command. 'Now. Go! You have to, Sally.' He gives me a gentle push and then I feel my feet moving beneath me before I have a chance to object.

I run back through the woods, the same way I came less than an hour ago, the wind once again whipping through my loosened plait, the undergrowth shredding my once-perfect dress. By the time I reach the reception, I am panting, tears falling down my face, ruining my bridal make-up. The party comes to an abrupt halt as they see me, confusion, then fear staring back at me in hundreds of different eyes.

'Help!' I scream through the room. 'Someone, help me!'

CHAPTER FORTY-FIVE

I sit with a blanket around my shoulders, shivering despite the warmth of the day. Someone has brought me a drink of water in a plastic cup, and I look back into the reception room full of discarded champagne flutes and half-drunk wine bottles. What a waste.

Our entire wedding party is outside on the lawn, not posing for pictures and toasting our marriage, but being interviewed by the mass of police officers that have descended on the scene.

Nathan has been taken away with Amanda. This was always the plan and yet I feel abandoned without them – the one left behind once again. Alone. But it's what we agreed. The instructions were simple.

I call for help; they stay at the scene.

I say nothing. Too shocked to explain what has happened in the woods.

The police find Natalie's broken body.

Amanda gives herself up.

She will be arrested.

What will happen after that? I have no idea. But it will be nothing good.

'Sally?' Chloe sits down next to me, her face scrunched up in concern. 'Are you okay?'

I nod, which is stupid because of course I am not okay.

'I can't believe she came back for Abi,' Chloe says.

I don't say anything, can't say anything because Chloe is angry and disgusted at Natalie – but would she be if she knew the truth of what Amanda did to her? If she knew her child was at the centre of years of hurt and lies?

'I'm so glad you were there,' Chloe whispers, her hand on her chest. 'If you hadn't been, I don't know what would have happened. She would have taken her, wouldn't she? She'd have taken her for good.' Chloe cries now, openly, the agony at the idea of losing her little girl flowing out. Chloe, who has done nothing except love a child that came to her by chance – a child that did not belong to her at first but now belongs in a way that is undeniably right and true.

I see Natalie's body on the forest floor in my mind. Feel the impact of the rock as it cracked against her hard skull. Did I go into those woods knowing that would be the outcome? What else could have happened?

There was never going to be a happy ending for Natalie.

Girls like Natalie never get their happy endings.

EPILOGUE

SALLY

ONE YEAR LATER

The wind whips through my hair as I walk across the beach and I scramble to find my hair tie. Once I've scraped my hair back off my face into a tight ponytail, I lift my sunglasses from my eyes and there she is.

She walks up the beach, cigarette in hand, smiling and waving at me.

I tut, inwardly, at the cigarette but there's a swell of pride at her appearance otherwise. She looks healthy. Happy. She's put some more meat on her bones since I last saw her and that under-fed look is almost completely gone.

'Sally!' she calls as she reaches me, before stubbing out her cigarette on the wet sand.

'Tiff,' I say, pulling her in for a hug, the word still unfamiliar on my tongue. 'How are you doing?'

'Good,' she says. 'I'm actually really, really good.'

–

We've come to the cafe at the end of the beach and it's full of sand-covered dogs, families and couples on a

Sunday morning post-beach-walk caffeine hunt. Without the wind whipping around us, the day is warm, and I take off my jacket and sling it over the back of the chair. Tiff does the same.

'Bet you're well nervous for today, ain't you?' she asks, and I nod.

I've needed Tiff's blunt manner this past year; everyone else has tiptoed around me but this girl wouldn't have a clue how to. It's been refreshing, and the fact that you can't really shock her has helped. She reached out to me once the whole story came out – the papers went wild covering it and to Tiff it was exciting that she'd been involved in the scandal – and we've kept in touch ever since.

It was hard to keep my friendships from the police while my family was at the centre of one of the biggest stories in the media. The story went viral because it had everything – undercover policing, corruption, teenage pregnancy, county lines drug trafficking and, to top it all off, murder. I doubt there's a person in the entire country who hasn't heard of Natalie Pierce by now, and of course, Amanda and Nathan Yates. My name was there too, but in the margins. I was not a person of note, except occasionally mentioned as Nathan's weeping bride.

I twist my wedding ring around my finger, wondering if today is the day I'll finally call my husband back.

'You'll be alright,' Tiff says, after our waitress brings over our drinks – flat white for me, hot chocolate with all the toppings for her. She smiles, a glob of cream on her nose, and I try to believe that it could be that simple. 'Sure she ain't still gonna be mad at you. It's been ages!'

'People don't always get over stuff that quickly,' I tell her now and she shrugs, then looks out of the window.

'World would be better if they did,' Tiff says, with that strange blip of wisdom she sometimes comes out with, before taking a loud slurp of her hot chocolate. 'We'll need to go soon though, visiting hours are only until three.'

—

'You came,' she says, her voice hoarse, like we're the first people she's spoken to for a long time.

She's thinner than when I last saw her, older, and guilt creeps through my veins that she's spent an entire year of her life in here alone, locked up, because of me.

'I didn't think you would,' she says, and I scratch the back of my head, not wanting to say that I didn't want to, that it was Tiff who talked me into it. Said it would be 'healing', something she'd no doubt heard on TikTok.

'I made her,' Tiff says with a childish smile that breaks the tension. 'I need a ciggie, back in a bit.' She flashes me a quick smile then exits the room before I can object.

Now, it's just the two of us, silent except for the ominous beeping of the machines.

'So, how are you, really?' I ask, pulling at the hem of my jumper to avoid her gaze.

Natalie has been in a coma since that night – the night I thought I had killed her. It was her breathing that fooled us – none of us could feel her breath, but it turns out she was breathing all along, left bleeding in the dirt and dust from the forest floor as I ran to get help. I've thought so many times how lucky Natalie was that Amanda believed her to be dead.

Amanda was arrested and is still waiting on a court hearing. The whole story came out, eventually. Minus the part where I attacked Natalie. Amanda held her word on that one and took the blame for me. It's something that doesn't sit easy with me, and in the middle of the night I still have to stop myself from picking up the phone and calling it in myself. Nathan was suspended and arrested for his involvement in what happened to Shane that night, although not for his murder. Amanda will face the penalty for that, as she should.

Even if he isn't formally charged with a criminal offence, Nathan will never be able to go back to the police. In some ways, this is the worst punishment he could face, and I have no idea what he'll do with the rest of his life. I left the police by choice and have escaped the entire saga without any repercussion, and that isn't fair – or right.

I've been anxiously waiting for Natalie to wake up – to see what she remembers, what she'll tell the police – and that moment has finally come.

'The police said Amanda confessed to everything,' Natalie says now.

'Yes,' I say, picking at the skin around my nails.

'She killed Shane,' Natalie says, and I nod. 'She blackmailed me into disappearing. She made my flat look like I was neglecting my baby.' I nod, again. 'And she attacked me at your wedding.'

Natalie stares at me, her eyes large and accusing. I don't nod, can't nod, can't move.

'I'm sorry,' I whisper.

Natalie breathes out, loudly, a crackle in her airways turning into a cough. 'Oh, Sally…' she says, and I look up at her. 'I was hoping I'd remembered it wrong.'

I shake my head. 'I'm so sorry.'

Natalie nods sadly, then rubs her face with the side of her hand, leaving creases by her eyes. She's aged ten years – I guess a traumatic head injury will do that to a person.

'I don't even really remember that day...' she says, her brows furrowed. 'I didn't mean to scare Grace.'

'I know,' I say, remembering the fear on Abigail's face that day in the woods. How she's seen things no little girl should have to see.

'I just wanted her back,' Natalie says, her voice breaking as tears start to fall down her face. 'But I've fucked it up, haven't I? They'll never let me see her now.'

I shake my head, sadly, because Natalie is right. No one in their right minds will ever let her near that child again.

'Maybe they will one day,' I say because it's the only kindness I can give her – some hope.

But she shakes her head. 'Nah,' she says. 'They've already told me there's no chance. They say I'm a "danger" to her.' She makes air quotes around the word 'danger' and it makes her look younger, a small child buried in a hospital gown, helpless and alone. 'Probably for the best,' she says and looks down, trying to save face. 'Still, they said it's unlikely the police will pursue the abduction charges.'

'Really?'

She nods. 'Looks a bit harsh, don't it? After what they did to me.'

'It wasn't the police though,' I say, old loyalties weighing heavy on me even now. 'It was just Amanda.'

Natalie frowns, like this distinction doesn't matter.

'And your dad,' she says. 'You can't pretend he weren't involved.'

I take in a sharp breath, avoiding her eyes. I was hoping – stupidly – this part of the story would not be true. That my dad wasn't involved, but of course he was. I need to face up to it.

'I'm so sorry, Natalie. If I'd known…'

She sighs. 'He weren't bad to me you know,' she says. 'He looked after me.'

'What?'

'He used to come and see me. We'd always meet on the beach; he said the sea air would do me good.' She laughs, not unkindly and I frown, trying to understand. 'We'd talk, walk for ages, sometimes. He'd bring me nice things – as well as the money – books and that. Didn't really read them but still, it was nice. Like he cared.'

I frown. 'My dad, he spent time with you?'

She nods. 'He was alright. You could tell he felt bad about it – all of it – like he didn't really want to make me stay away. He told me about your mum, you know. All the drugs and that. I think he felt guilty about making her leave you, and I was the only one he had to talk to.'

'I don't understand why he did it. Any of it,' I say.

'I think he did it for you,' Natalie says.

'For me?'

She sighs. 'He said your mum would have ruined your life if he'd let her stay. Said she turned everything she touched to shit.'

I blanch at the callous words, unable to see how this was possibly him doing the right thing by me.

'He was just trying to keep you safe,' she says. 'That *woman*,' she says it with a sneer, now switching to talk about Amanda, 'and Nathan, they're the only family you've got. Ain't they? I guess your dad knew if he didn't keep me away like he kept your mum away, I'd ruin it all. Amanda and Nathan – they'd both go to prison and you'd be left alone. He was dying, weren't he? So, you'd have no one.'

She says it with such assurance – such simplicity – that it feels like the only answer. My dad wanted to protect me – in his own misguided, stupid way – he must have thought it was the right thing to do. Just like he believed forcing my mum out of my life was the right thing, too.

'Why did he change his mind?' I ask her. 'Before he died, he told me he didn't think I should be with Nathan.'

She shrugs. 'People see more clearly when they're about to go, don't they? Like me. I always thought I had to get Grace back to ever stand any chance of happiness, but I know now I need to let her live her own life, be happy – without me. I had my chance at happiness – I didn't keep her safe. I should never have got her involved with Shane that night. What happened after, yeah, that woman set me up, but I still put us both in that situation. Gracie's better off with that family. The nice one, with the big back garden.'

I start to tell Natalie that Chloe and Matt don't have a big garden, but then realise that's not what matters. She's told herself a story about Abigail – one she needs to believe to let go of her little girl – and I should let her have that. The truth won't always set you free.

'I'm never going to tell them, you know,' Natalie says.

'Tell them what?'

'That it was you that hit me.'

I shake my head, unsure what to say.

'Why?'

She shrugs. 'Figure you've been through enough. And that woman has a lot of making up to do – she fucked your life right up. Plus, your dad was always nice to me, so you know. I'd like to repay the favour. To do something good, for once – to be the person I promised myself I would become when I took your name.'

I nod – words stuck in my throat at this momentous gesture.

The door springs open and Tiff's head pops through.

'Nurse says visiting time is over,' she says, rolling her eyes before shutting the door again. I look over at Natalie. She smiles, and nods, as if giving me the okay to leave.

As I reach the door, I turn back to her.

'Will I come and see you again?' I ask.

She shakes her head, slowly. Not unkindly.

I nod.

'I've got my own life to live. So have you,' she says, her head resting gently on her pillow. 'Goodbye, Sally Jones.'

As I walk out of the hospital, Tiff by my side, I wonder what that life looks like. Right now, I have no idea. My dad's worst fears have come true – I've lost the family he was always so desperate for me to keep. But I know one thing for certain – there are scarier things in this world than being alone and it's about time I started putting myself first.

A Letter from Sophie

Dear reader,

Firstly, I'm terrible for including spoilers in my author letter and acknowledgements, so please only read these once you've read the book!

Secondly, thank you. The world is increasingly fighting for your attention – TV is *so* good and TikTok is unbearably addictive – so it's a wonder we ever manage to read books at all. But we do – I do – you do. Thank you for reading mine. Whether this is your first book from me that you've read or you've been here since book one, I hope you've enjoyed *The Other Miss Jones*.

About ten years ago, I 'met' the other Sophie Flynn when she received an email meant for me and kindly forwarded it over. Since then, she's received emails about my pension, mortgage, random appointments, and much more. She knows an awful lot about my life, and because I'm a writer, this funny situation sparked the idea for a book.

It took years before I finally managed to turn that spark into a full book. I kept trying, but it never quite worked – until I wrote a short pitch and sent it to my publisher. They immediately wanted me to write the whole book, so I had no choice but to actually do it.

It has not been easy! If you're reading this and thinking of writing a book, my biggest advice would be do NOT give the characters the same name and have the crux of the story rest on this fact! You will slowly lose your mind this way and it will require more spreadsheets than you are comfortable with in order to keep track of things.

But anyway, I did eventually get there and have since learned that many people have this same email relationship with their name-twin (a term my editor would not let me use in relation to the book as it reminded her too much of Joey's hand twin from *Friends* – fair enough). If you do have a name-twin, please do send them this book! And the other Sophie Flynn, if you're reading this, I really hope you enjoyed this book. I don't think you're a criminal who took my name and pretended to be someone else, I promise.

As well as writing about my own strange email situation, I've always wanted to write police characters and so I hope you enjoyed seeing a small insight into life as a call handler. My sister did this job for many years and for a short time we worked for the same police force – though I 'fannied about with the press releases' (as Daniel Cleaver would say) and was not responsible for making life-and-death decisions.

The police 'family' is called that for a reason and I've always found it a very interesting concept – the idea that your home life and work life are so connected. It can be a dangerous thing. Of course, for most people in this situation, things work out just fine. But those stories don't make books and so I wanted to explore what would happen to a family who did something 'bad' when their whole lives are built on being 'the good guys'.

One of the things I struggled with in this book was giving it a satisfying and true ending. It hurt my heart that Natalie would not end up reunited with her daughter, but there really was no way to give a happy ending here and I don't doubt that some of you will find that unsettling. I like to think that one day, Natalie would be able to have contact with Abigail and that, regardless, Abigail would grow up knowing she was loved by everyone around her – even though that love often caused them to do very bad things.

I will finish this off by saying thank you once again for spending your precious time reading my book. I really hope you enjoyed it and I would love to hear from you if so! The best way to support authors is by leaving reviews on places like Amazon and Goodreads. I've been told I'm not supposed to read my reviews, but I always do. To keep up with my writing life, do connect with me on Twitter @sophielflynn and Instagram @sophieflynnauthor, where I'm always thrilled to receive a message from readers.

Thank you again,

Sophie x

Acknowledgements

I will start, as always, by thanking my wonderful editor, Jennie, and publisher, Hera. This is my fourth book with Hera and I feel so lucky to be published by such a hard-working and brilliant team. There are so many people involved in the making of a book that often go unseen – so please let me take a moment to name everyone who has been involved in making mine: Jennie Ayres, Keshini Naidoo, Dan O'Brien, Kate Shepherd, Lindsay Harrad for the copyedit, Vicki Vrint for the proofread and Mark Swan at Kid Ethic for the cover design.

Booksellers across the country, thank you! Spending hours milling around bookshops has always been one of my favourite activities and it is greatly enhanced now by spotting my own books on the shelves. A particular thank you to Borzoi Books in Stow, for always stocking my books on publication day so that I can bring in my family and take a photo with the gorgeous Prince (the cutest bookshop dog you have ever seen), and Waterstones in Cheltenham, where I've had the joy of seeing my books on the table multiple times. Alongside booksellers, the book blogging community in the UK is pretty wonderful and always so supportive of my books, so thank you.

Thank you to the professionals who helped me with research for this book. In particular, Graham Bartlett who advised me on policing matters with kindness and patience! Also, Emily Kelley from Adoption Central England, who very calmly answered my mad Friday email rambling on about 'what would happen if a child was adopted because their birth mum was set up for murder but she turned out to be innocent'. Emily, it appears, is rather unshockable.

Thank you to everyone in the OCC at Warwickshire Police, who are so brilliantly supportive of my books – I hope my references were accurate, though I'm sure many of Sally's actions had you shrieking 'she could never do that!'. Any mistakes in the processes are, of course, mine and mine alone (though OCC readers, you can partially blame Rosie, if you like).

I thoroughly enjoyed my three-day DIY writing retreat at the Premier Inn in Weston and hope those that know the town well think I captured it fairly. Like Sally, I used to be taken on trips to the seaside as a kid and loved the arcades and fish and chips – and those remain my favourite things.

Writing a book can be quite lonely and so I'm very thankful to my brilliant writing friends who are never more than a voicenote away. Particular thanks go to Liv Matthews, Meera Shah, Becca Day and Sarah Clarke for their endless words of wisdom, and a special thank you to Bev Walsh – you may not remember, but while we were queuing for coffee one morning, you managed to steer me away from a bonkers plot idea that would have caused me no end of issues. Thank you also to all the authors who read an early copy of *The Other Miss Jones*.

My colleagues at Jericho Writers – I couldn't work with better people and you all make writing a book while running our small company a whole lot easier.

2024 has been a rollercoaster year and I've never been more grateful to my friends and family who have provided unlimited support in so many different ways. Particular shout-out to Natasha, Anna, Rachael, George, Janelle, Pearl, Verity, Alex and Zoe.

My family: Mum, yes, I've finally finished editing! You no longer have to ask me every week if I'm still 'doing my book'. Dad, thank you for the copies that you continue to flog to everyone you meet. Rosie, you've already had your dedication, but thank you for answering endless police questions over the years and providing inspiration for Sally Jones. Nicky, for always being on hand to swap books with. Jill, Steve, Dan, Mieko, Hibiki and Hikali for being brilliant, as always.

Tom, thank you for always making me pasta when I'm working late on my edits, waking me up every day with a frothy coffee, and never, ever saying the words, 'Don't you have enough books?' I couldn't do any of this without you.

And finally, you, my lovely reader. Without you, these books don't exist. Thank you.